Simon Cleary is the author of three novels, including *The Comfort of Figs* (2008), which was published after the manuscript was shortlisted for the Queensland Premier's Literary Awards. His second novel, *Closer to Stone* (2012), was inspired by his experiences in North Africa at the commencement of the Algerian civil war in the 1990s. It went on to win the Queensland Literary Awards People's Choice Award in 2012. Simon's third novel, *The War Artist*, was published in 2019. *Everything is Water* is his first work of nonfiction. He is a lifelong walking and environmental enthusiast, and lives in Brisbane.

EVERYTHING
IS
WATER

Simon Cleary

First published 2024 by University of Queensland Press
PO Box 6042, St Lucia, Queensland 4067 Australia

University of Queensland Press (UQP) acknowledges the Traditional Owners and their
custodianship of the lands on which UQP operates. We pay our respects to their Ancestors and
their descendants, who continue cultural and spiritual connections to Country. We recognise
their valuable contributions to Australian and global society.

uqp.com.au
reception@uqp.com.au

Cover design by Christa Moffitt // Christabella Designs
Cover photograph by Bibadash // Shutterstock
Author photograph by Patrick Hamilton
Typeset in 12/16 pt Adobe Garamond Pro by Post Pre-press Group, Brisbane
Printed in Australia by McPherson's Printing Group

University of Queensland Press is supported by the
Queensland Government through Arts Queensland.

University of Queensland Press is assisted by the
Australian Government through Creative Australia, its
principal arts investment and advisory body.

A catalogue record for this book is available from the National Library of Australia.

ISBN 978 0 7022 6850 2 (pbk)
ISBN 978 0 7022 6979 0 (epdf)
ISBN 978 0 7022 6980 6 (epub)

University of Queensland Press uses papers that are natural, renewable and recyclable products
made from wood grown in well-managed forests and other controlled sources. The logging and
manufacturing processes conform to the environmental regulations of the country of origin.

In memory of my father,
an inspired walker

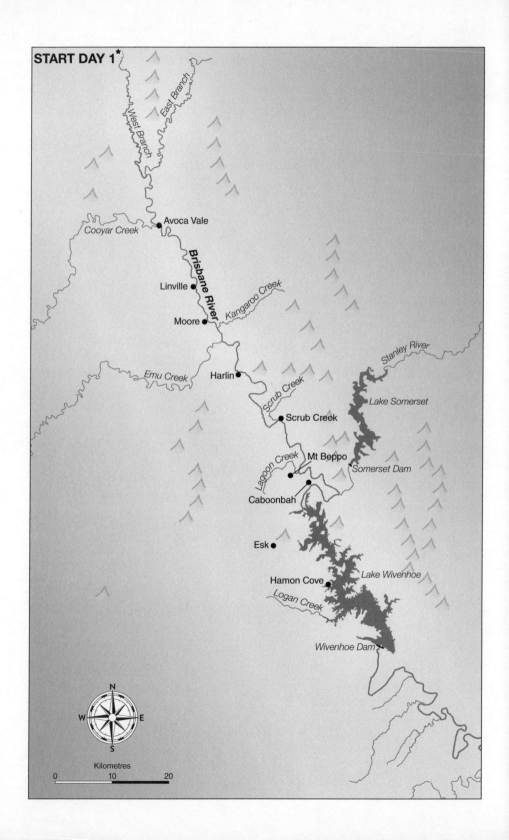

START DAY 1

West Branch

East Branch

Cooyar Creek

Avoca Vale

Brisbane River

Linville

Kangaroo Creek

Moore

Emu Creek

Harlin

Scrub Creek

Stanley River

Lake Somerset

Scrub Creek

Lagoon Creek

Mt Beppo

Somerset Dam

Caboonbah

Esk

Lake Wivenhoe

Hamon Cove

Logan Creek

Wivenhoe Dam

N
W   E
S

Kilometres
0   10   20

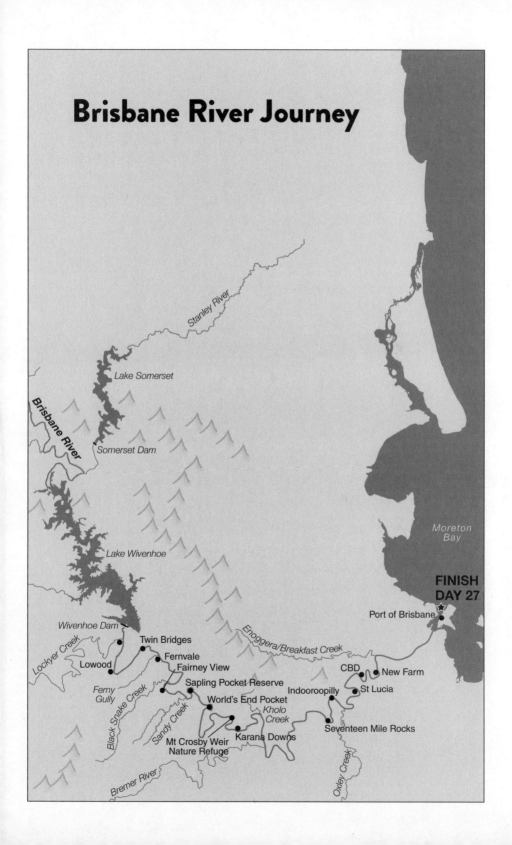

Brisbane River Journey

Stanley River

Lake Somerset

Brisbane River

Somerset Dam

Lake Wivenhoe

Moreton Bay

FINISH DAY 27

Port of Brisbane

Wivenhoe Dam

Lockyer Creek

Twin Bridges

Lowood

Fernvale

Fairney View

Ferny Gully

Black Snake Creek

Sapling Pocket Reserve

Enoggera/Breakfast Creek

CBD

New Farm

Indooroopilly

St Lucia

World's End Pocket

Kholo Creek

Sandy Creek

Mt Crosby Weir
Nature Refuge

Karana Downs

Seventeen Mile Rocks

Bremer River

Oxley Creek

Beginnings

There is an old story. It has the power of myth so might yet, I tell myself, become true again. A river rises in the mountains, gathers water from its tributaries as it descends, snakes across its flood plain before spilling into the sea. Creating. Destroying. Altering. Cleansing. The river obeys a natural law and pursues an ancient destiny: to flow. In that story the river is more than just a source of water.

There is a river. It is the same as all rivers, follows the same ancient laws. But it is also unique, as all rivers are. It is *my* river. A geographer might say of my river that it is 344 kilometres long and sinuous. A demographer, that it sustains a series of small settlements in its upper and middle sections and a rapidly growing capital city near its mouth. A hydrologist might be satisfied that the dam blocking its path stores the water the city needs and reduces the risk of the city flooding. A water quality scientist might say that things could be worse and have on the whole been improving. A canoeist might celebrate such a long waterway so close to home. And a poet might say, simply, that the river is many rivers.

There is a boy who used to play in the creeks that formed at the top of the Great Dividing Range. He is no different from any child who's ever gone down to a creek to explore, his sense of wonder too great to contain. One day, aged ten or eleven, he follows his favourite creek down the escarpment. Beyond the tadpole pools, past the rockslide and the gnarled fig tree, and down the side of the waterfall that marks the bounds of his childhood territory. All morning he walks, following his nameless creek wherever it leads till he reaches the bottom of the range and makes his way through scrub and across paddocks to the village by the highway and calls his father from the public phone booth to collect him. The boy doesn't know that his creek feeds Lockyer Creek that in turn feeds the Brisbane River that flows through the capital city near the coast. But the boy knows other things. He knows that his creek is a tangible part of his world, solid, real. That you drink from it, head bent like a wallaby. That it is, in fact, at the centre of the life of the bush around him. That the frogs and the maidenhair ferns and the lantana and the lizards all depend on it. That whether it flows or not depends on the rains, and that while it might run dry, the rains will come and it will flow again. That it can be depended on. And that a kid can safely fix his dreams to it.

~

So, having settled in Brisbane, the city on the river known as Maiwar by some of the First Peoples of this country, it is to the river I turn. Like the city, my life here is unimaginable without the river. Yet what do I know of it? Really? I know my childhood creek, and I'm familiar with those reaches of the river that pass through the city. But upriver there are hundreds of kilometres I am oblivious to. And if I am ignorant of so much of the river's existence, how much of myself can I not see?

A seeping idea. How long it might have travelled the currents of my subconscious I can't say, but swim to the surface it surely did: following the river for the entirety of its length might help me understand.

But how to follow it? From sea to source or the other way round? There is an attraction to starting at the bay and tracing the river to its beginning. That is the way of understanding a thing by going back to its origins. But an upriver route from the known into the unknown is also the way of explorers and merchants and colonial history. It is the way of Sir Walter Raleigh navigating the Orinoco in search of El Dorado's golden riches. Of Livingstone and Burton and Speke and their nineteenth-century expeditions in pursuit of the source of the Nile. It is also the way of the handful of British men of the 1820s who successively pressed further upstream in the first decade of this river's colonisation. I have no wish, now, to follow in footsteps such as theirs.

No, what I want is to begin in the mountains, at its first trickle. I want to follow the river from its source in the Great Dividing Range to the coast. I want to follow the river I love for however long it will take, from the quiet of the mountains to the busyness of the metropolis, from solitude to community.

I want to go *with* the river's flow, not against it. I want to follow where the river leads, to listen, to observe, hopefully to learn. And I want to *walk* the river's journey. To walk beside it. On its banks, travelling through the landscapes and towns that are watered by it, rather than going by watercraft and seeing those places pass by. I want to walk with the river as companion. I want to go slowly, carefully, attentively. To walk as a pilgrim might walk. To submit to the journey, whatever its gifts, whatever its obstacles.

The rhythm of the journey itself will guide me. I will lose myself in the walking and its meditative pulse. To see anew. If need be, to unsee.

Part of me wants to travel sparely, with boots and a simple pack: a mat to lay upon the ground at night, a blanket against the cold, a second set of clothes should it rain. To go without a map, because how could one get lost following a river? Without phone, because all the world I need is around me. Without even a store of food, setting out like a mendicant, trusting to the universe. What might I offer the universe in return? Not prayers, as wandering monks once did, but an accumulating story. Of the river and its wonders.

But the world is not so simple. It is infinitely more interesting. If I was to walk beside the river it would mean walking through history. Recent and ancient. It would mean walking through people's properties. First Nations' lands. Farmlands. Crown land. The lands of people I'd need to ask permission from. People who lived intimately with the river and knew its ways better than me, better than I ever could. People who might also be willing to share some of what they know.

I started with two men whose families had lived on the river for generations: Graham was an old mate of my father-in-law, Andrew a friend of both a schoolmate and one of my sisters-in-law. I explained what I had in mind: to walk 340 kilometres – the length of the river over a month or so, sleeping on its banks, gathering stories of its people, its history, its ecologies and perhaps to write about it. They nodded, and whether they were yet won over, they were at least prepared to introduce me to their neighbours, who in turn gave me the names and phone numbers of theirs – a domino effect of introductions up and down the river. I reached out to Traditional Owners, to government agencies and community organisations – anyone with responsibility for the river. I had hundreds of conversations over eighteen months, some in person, others by phone. The people I spoke with were enthusiastic, interesting, generous with their stories. In time, I came to understand that somewhere in those many months of planning my river journey had already begun.

Rather than resisting the need to plan the walk, I needed to embrace it. Working out logistics was an act of imagination. Because no visible path existed along the length of the river, a walk like this needed to be imagined into being. I realised that I alone couldn't do that work – it could only be a collective act. Slowly, almost miraculously, the contours of the walk began to take shape. The journey began to gather fellow-walkers to it, people who were animated by the river, each keen to experience the river on foot: a Jagera man working to preserve cultural heritage, a professor with a specialty in sustainable water use, one of my two sons, my naturalist brother, one of my sisters, a flood engineer, a journalist who'd covered natural flooding disasters. With each new walking companion would come not only fresh supplies, but, I hoped, fresh perspectives on the ways of the river.

By my reckoning there were two stretches that couldn't be walked: a seven-kilometre reach where I'd need a canoe to bypass the city's cluster of prisons, and the final leg of the trip past the port then through the mouth and into the bay. That last stretch would have to be completed by boat. A friend offered his refurbished timber cruiser with its shallow draft, perfect for Moreton Bay, should I need it. By which he meant, should I *make* it. And it was true, the mouth of the river was a long way away.

~

Then, in February, just three months before I planned to start, the river flooded. Again. There were weeks of rain, torrential towards the end. Great volumes of water flowed over the land, rushing across the city's concrete and bitumen, far too much for the river and its tributaries to hold. Banks burst. Suburbs were inundated, streets cut off for days, homes destroyed, vegetation drowned, the banks themselves reshaped. Thirteen people died in this catchment, or in neighbouring ones. All around, chaos and destruction. And then, gradually, the floodwaters subsided.

The river does this every ten or thirty or fifty or eighty years or so. It stirs itself from its meandering languor, rises and spreads its swollen destruction over the land. It drowns and sweeps away and retreats and leaves its mess and stink. Some of us forget. For others there can be no forgetting, the losses too great. As for me, I fear I see too little. That I forget too much. That I fool myself about the quality of my love for this river. That I love too shallowly.

Should the walk proceed? I spoke to river people up the valley, and made reconnaissance trips to gauge the damage. Yes, the banks have been eroded in places, but, I heard again and again, the river is resilient and so are we.

~

On a new moon, in a new week, in a new month, I leave: Sunday 1 May. It's a good day to set out on a long journey, faintly auspicious. And May, in this part of the Southern Hemisphere, is usually an ideal time for walking. The heat has come out of the year, the summer rains are done, the snakes have begun to hibernate and there's still enough light to set up camp after a decent day's hiking.

The day itself, as I wake in the dark of my Brisbane home, breaks propitiously too. A night creature is scratching its way across the corrugated iron roof, clawing open whatever it was I was dreaming. The creature leaps from dream to gutter to tree and is gone. I reach for the dream, but it too has gone. My heart pounds. What I hear next is the sound of water from overnight rain, dripping into the water tank outside the window. I hadn't heard the rain through the night, though it had been forecast. Now, the steady dripping is soothing. My first conscious thought – quotidian, practical – is whether it'll be raining when we get to the top of the range today. If so, I'll need to set out on the walk wearing the rain jacket I'd packed but hoped not to have to use. I check my bedside clock, but it is too early to rise. There is time yet to listen to the dripping water and to contemplate these droplets – the very reason for my walk – and how mysterious their journey is. Where have they come from? How far have they travelled? What have they seen? Soon enough they'll make their way into the catchment to join my river, a river that has been flowing for millions of years. Surely these droplets falling in the pre-dawn dark outside my bedroom window are a benediction, and this month-long pilgrimage along the banks of the river from its source up on the range to the city and then out to the salty bay is blessed. Surely.

Day 1

Source

Distance: 14.76 km

Evening Camp: Place where rain falls lightly on laughing water

We will start at the source, whatever that may be. Geographers and cartographers need a river to have a source. But settling on one isn't always easy. A river has multiple tributaries. How to choose which is the trunk – the start of which will be designated the river's source? Is it the longest tributary? The one with the greatest catchment? The one carrying the largest volume, or that with the highest elevation?

The geographers and cartographers and surveyors are forced to choose. But for this river the choice is too hard. There are two tributaries, each rising on the escarpment of the Great Dividing Range, about 140 kilometres, as the crow flies, northwest of Brisbane. The first tributary flows around one side of a small mountain, Mount Stanley, and the second flows round the other side, till they join just south of it. The measuring people say, let there be two branches of the Brisbane River, and they name them so: the Western Branch and the Eastern Branch. They ignore the claims of downstream Cooyar Creek, with its greater catchment, and Lockyer Creek, the longest of the tributaries. None of the tributaries care.

So two branches. From the start of which, then, does one begin measuring? And from which point exactly? And how does one factor in, when measuring, the curvature of the earth and the drop of waterfalls and the passage of a river when it splits around an island? To say nothing of the changes over time of a sinuous river's course? The river doesn't care. The books say this river is 344 kilometres long, so let's accept that, but let's also accept it's an estimate, and that when the next flood gouges away a bank or bursts through an inside bend, it will be wrong.

So let this be a 344-kilometre walk along the river's banks, sleeping each night by its side, within sight or sound of it.

I choose the start of the Western Branch, or rather, it is chosen for me. An old riverman I've come to know says the river starts on a property up there – a farm owned by his cousin, high, windswept, scoured by a hundred gullies. That's where the source is. Let me take you.

~

The riverman, Graham, is in his late seventies and has worked properties and machinery all his life. His home is in a village, Moore, at the foot of the range, though he also has a farm on the river nearby. He is from a 'river family', a term I'd heard often enough while planning the walk. There are river families all through this country: families who've lived on the river for three, four or five generations. The oldest families trace their river-links back to the nineteenth century, and one or two to when the Jinibara and Jagera were forced off their land in the 1840s. That's a while, but it shrinks when set beside how long the river's First Peoples lived here. And is the mere blink of a geological eye when one thinks of the age of the river itself. Graham's ancestors, like half of mine, come from Ireland, though while mine were Catholic, his were Protestant. His family mythology has his forebears emigrating to Australia in time to join Burke and Wills on their doomed expedition in 1860. He has a blood memory of failed expeditions. He's anxious for me.

So those of us starting out on the first leg of the walk assemble at Graham's place. Steve, a professor of water and environmental biotechnology from one of the city's universities, will walk with me for the first five days, from the source to the old timber town of Linville. Our paths first crossed when we were kids, in Toowoomba, and we've only now reconnected, years later, over the river. Today we're joined by James, a Jagera man who's worked on Aboriginal archaeological sites throughout his Country. Unlike us, James knows the river and many of its downstream tributaries intimately, knows them in ways Steve and I never could. He's never, however, been to the river's source. My wife, Alisa, comes to see us off, and so does Steve's wife, Nic.

When Alisa and I pull up at Graham's we're met with two quite different signs. Graham has planted a huge corflute placard just inside his fence, announcing his support for the local conservative rural-interests political candidate in the approaching federal election. On the other side of the fence, parked on the grass verge, is James' LandCruiser. The large lettering on the decal on either side of his vehicle is equally prominent: *Jagera Daran Community & Heritage Solutions*. Different sides of a fence, different signs signalling different histories and political cultures.

Steve and Nic soon arrive; we all introduce each other, find common ground in the excitement of starting out together on a shared experience, and take opportunities to laugh at our own expense. There's a sense, even before we've begun – before we've even reached the river – that this is a time to lay our egos aside and to make ourselves small.

~

We follow Graham in convoy up the river road as it threads the upper reaches of the valley, climbing the range to the watershed. The road's relationship with the river feels, at first, strange, and uncommitted: following it briefly before turning away, again and again, to find a more direct route. Until I realise the road's loyalties are different. While the river faithfully follows the contours of the valley – indeed, has created the valley itself – the road's interest is only in ploughing forward, up and over the range. For this doggedly determined 'Western Branch Road', the river is an obstacle to be overcome.

The convoy stops on occasion as we make our way up the range. We all get out of our cars and walk up to Graham's, so he can point out landmarks. He loves this land, is proud of it, wants to share it. We stop where the road crosses the river at the first of the thirty-eight causeways we'll need to traverse as we scale the range. We stop near a set of old split-rail cattle yards where Graham worked as a kid. We stop near 'Stony Pinch', where Steve and I hope to camp tonight on our way back down the riverbank, for Graham to point out what our best approach would be. We stop where the road falls away to the left and a beautiful view opens up across the thickly grassed valley, the river itself concealed by parallel lines of ti-trees on the narrow valley floor. We gather round Graham as he points towards a stand of ironbark on a far ridge. A discussion ensues about what ironbark can be used for. This, that, the railings of the stockyard back there. The talk peters out as each of us continues to gaze across the valley.

'What did it use to be like?' Steve asks. 'What did the country use to look like?'

By which he means, of course, before whites. It's the question we're all asking ourselves.

Graham answers. Carefully. Thoughtfully. It's a question he's obviously asked himself often. He can't say what it looked like before settlement, he tells us, but he knows what it was like when he was a kid. The ironbark was thicker on that ridge, he says, pointing. There were hoop pines all over the mountains. And the valley has been opened up a bit more, too.

'Might it one day return to how it was?' someone asks.

He seems to weigh a lifetime of experiences before answering with a sigh.

'What you've got to realise,' he says, 'is that this is cattle country.'

There's a pause, long enough for James to quietly intervene.

'It used to be kangaroo and emu country.'

Graham doesn't respond. No-one does.

We all know that, long before white settlement, in the Dreaming, Country was formed by Ancestor Spirits. We learnt it at school, or from Indigenous paintings, or from First Nations Elders welcoming us to Country. We've all heard, or think we have, how a Dreaming Rainbow Serpent once travelled along this valley, slithering all the way to the bay, gouging out the river's course as it went, its great track becoming its bed and banks. Though perhaps it was a variation of that story we learnt. Perhaps the river was a carpet snake, or bunyip, or eel. Landscape shaped in the Dreaming. Powerful ancient stories of creation told by the Jinibara, Jagera, Ugarapul, Turrbal and Quandamooka people through whose Nations the river passes.

Today the river is known by many of the city's people as 'the Brown Snake'.

What James doesn't tell us, not here, not now, is another truth: that one of his near ancestors, Jung Jung, was Emu skin, an initiated Jagera man who wore the scarification marks of his Emu clan.

We return to our cars and resume the drive up the range, the track of a Dreaming Rainbow Serpent to our left. As I look out the windscreen, I think I can see, surrounding me, kangaroos and emus. They are everywhere. I see eastern greys lounging in the shade where they have been resting for millennia, joeys crawling out of pouches, roos bending to drink from the ancient river, bounding across the eons towards me. I see emu chicks hatching from eggs laid on a thatched mound of twigs on a sandy beach on a bend in the stream, breaking free, striped with beautiful history. I watch them grow before my eyes, and race, all that swaying shaggy plumage, those long necks stretching forward. Racing. Outracing. Blink. Gone.

As we near the very top of the range, we enter a mist-shrouded forest and a new ecosystem. From every damp gum hangs the silvery-green tassels of old man's beard, *Tillandsia usneoides*, rootless, an introduced species, now naturalised. The forest is so thick with the cascading strands it could almost be an everglade we're passing through. I stop the car and wind down the window. The silence is eerie. The shimmering epiphytes hanging from their host trees in the fog could be spectres. But whose ghosts? And is this welcome or warning?

Graham pulls off the road at the entrance to a property on the left. We follow him up the dirt track to the gate. Graham gets out to open the gate. His wife moves across to the driver's seat and rumbles over the grid. Graham holds the gate open for each vehicle to pass, Alisa's and mine, Steve and Nic's, and finally James'. As is the convention in the bush, James stops when he's through and waits for Graham to close the gate. As is also the convention, Graham walks around to the passenger side of James' vehicle and opens the door – *Jagera Daran Community & Heritage Solutions* – and sits beside James. Cattle country. Kangaroo and emu country.

We drive in slow single file following a high ridge until it becomes a finger of land, until there is no land anymore and ahead of us is a huddle of sheds and vehicles. The country falls away on three sides, the slopes well grassed after a good season. Waiting to welcome us, hands on her hips, is Graham's cousin Marjorie. She is eighty, slim, fit, curious, strong, hospitable. She, too, is a river person, but this property is unique. Not only does the Western Branch of the Brisbane River rise here, but some strands of local lore have it that the Mary and the Burnett rivers do too. That different gully systems on the property feed into three different catchments.

We all get out of our cars for introductions. With Marjorie are her husband, one of their sons and some of his mates who've gathered for an annual camping weekend. But I'm distracted. My eye is on the landscape: the ridges and the hills, the highest point of land, the folds in the earth. I think I can identify the gully we're here to find, but the low, grey sky makes reading the contours of the country harder, and there are so many gullies carving their slow way through the hills all around. I resist the urge to ask Marjorie, too early, and join everyone for morning tea.

A dozen or so of us mill around the simple hut and its old companion creamery – the only buildings remaining of what had once been a dairy – as we chat over tea and scones. Welcomes don't come warmer than this. In time, Marjorie takes me by the elbow and guides me away from the crowd. She points out the gully we've come for, a fold in the earth to the left.

The river.

~

Scottish poet Nan Shepherd once observed that we can't know a river until we've seen it at its source.

The source of this river is unobtrusive. There is no spring bubbling from a subterranean reservoir. No melting glacier. No marsh filled with ancient seep. Up here, the water first entering the river is rain freshly fallen at the top of the range, either running overland with gravity, or leaching a little more slowly through the ground's uppermost soils. The gully receives what the sky gives. In those years the sky has nothing to offer, the river gully is completely dry for kilometres. So dry the imagination cannot even conceive of a thing such as water. But when the clouds that strike the range from the east are heavy, the gully fills with rain. It rained heavily in February and has rained often since. It is difficult to imagine a time before rain.

The range – the Great Dividing Range – is not a high range, not by the standards of other continents. And it's not particularly high here. I check my GPS. We're at 434 metres. Nothing. But the ways of gravity are slow and subtle. My father used to tell us, when we were kids growing up on the escarpment of the range, that if you spat to the east, your spit would make its way into the little creek that tumbled off the escarpment into the Lockyer Valley, then into the Brisbane River and on into Moreton Bay and finally the Pacific Ocean – whereas if you turned round and spat westwards, your little gob of moisture would take a longer route to the ocean across four Australian states via the Condamine and Balonne before joining the Darling and then the Murray and flowing into the Great Australian Bight. There was majesty in the idea – that we humans are part of a grand and wondrous natural scheme – but it also reflected the genius and predictability of science, of the laws of gravity and geology and climate.

~

It's nearly 11 am. It is time to go. I am not calm. There's the excitement of setting out on a long-anticipated journey. But I feel something else: that something elemental in this place is unsettling me. I can barely contain myself and I have to get off the ridge. I have to see what's in the gully. I need to drink from the stream.

Steve and I climb into our backpacks, and James slings his satchel across his shoulder. I gather my walking sticks. We say our farewells. The river pulls me from Alisa's embrace. It's pulling harder than I know how to resist.

We take the shortest route, down the hill, wading through the grass. Soon enough our friends on the ridge disappear from sight. We climb over a fence and hurry towards the line of trees growing thick in the fold of land before us. Foliage like that means it's not just a gully ahead, but a creek. We hear the tinkling of rivulets falling over shaded rock. And then, sure enough, we see it, and in all of creation there has never been water clearer than this.

We bend and scoop handfuls to our mouths.

I'm a child again and this is my first creek at the back of my childhood home and suddenly I'm surrounded by magic and mystery – who knows where that little flow of water tinkling its way downstream wants to go and how many adventures might be had by its banks. I'm smiling like a fool. But so are Steve and James.

~

The pull of this little creek cannot be denied.

We walk with river she-oak and callistemon bent to the river from February's floods. Half-a-dozen grass trees look down on us from a cliff. James points out a bats-wing coral tree growing from a fissure in the rock. The bark was used to make shields. He pulls reeds from the edge of the bank, lifts them to his lips and sucks out their sweetness. 'Have some,' he offers. We fossick among the stones in the creek bed, turning them over, showing them to each other, wonder upon wonder. Here we find crayfish claws, smooth and blue; there, maidenhair fern. James has an amazing eye. A snake has shed its skin on a lichen-covered rock.

We step onto a shelf of dark grey basalt, hundreds of smaller, lighter-coloured pebbles set into it like jewels. Can two hours have passed already? We stop to eat, dangle our legs over the edge of the shelf and inhale the cool water.

Soon enough the river beckons once more and we rise. James finds a clump of lomandra downstream, tears off a leaf and starts chewing. Then, ahead, we're confronted by something familiar but entirely unexpected – a barbed wire fence across the river. After the flowing creek, after feeding our curiosities, after the marvels we've been sharing so enthusiastically, we're momentarily unsettled. After the initial shock comes laughter. As if the river could be fenced by a few strands of wire! There's enough room, close to the water, to duck beneath its lowest strand without needing to take off our packs. But on the other side of the fence, things are different. The bank has become a wide patch of bare earth, turned and muddy at the water's edge where hooved animals have come to drink. A well-worn path runs up the hill away from the creek, parallel to the fence. For a while, as we continue our downstream journey, the river's waters are milky from the disturbance. Indeed this is cattle country.

James takes the lomandra from his mouth and begins twisting it in his fingers. A bone protrudes from a section of bank. We quieten, and the three of us lean close. Steve scrapes away the earth. The chances of it being human are slim, infinitesimal. Still, he's careful. This is how archaeologists work. But it's a marsupial, the rib of a bandicoot or wallaby. Steve lays it tenderly on the grass above the bank.

We check the time. James needs to give himself enough daylight to follow the river back to his car. He knows the way, but this time it'll be uphill. He really should be heading back about now. But here he is, and here the river is, and who knows what mysteries might lie just ahead? He's got a bit more time, he says. We continue on. Bend after bend, rock pool after shaded rock pool. The sun is dropping. I'm growing worried for him. One more bend, he says. But after that one, there's another bend. And another after that, too.

Eventually James stops. He shows us the lomandra twine he's woven. You chew the leaf to break down the fibre structure so you can weave them. He hands me the twine and we embrace, then he turns and makes his way back upstream, stopping here and there, till the river bends him out of sight.

~

Before us, the river continues to spill off the range, following a route circumscribed by deepest time. A route with its genesis in the explosions of stars and the gathering of cosmic particles into planets, and much later, into continents. By elemental forces. By gods and volcanoes.

So the rocks are ancient, and the river carving its way through this valley just a little less so. Say ten million years, late Miocene. Ten million years of carving its signature. Ten million years of laying down the outline of this course. Onwards we go, Steve and I, walking a ten-million-year-old way.

We follow the river down the range. Mountain and rock and water riffling over corrugated beds. Thick bands of callistemon, river she-oaks as large as any I've seen, girths of a metre and a half. A mickey bird crosses the river just in front of us, close enough for us to follow its dipping path. It lands on the lowest branch of a great tree, heavy with huge burls, dark and fibrous: a river red gum, dying or being reborn.

~

Suddenly it's dusk, so immersed have we been in exploring, so much has the river whetted our hunger to understand, so greatly have we enjoyed each other's company. It's time to select a camp site: flat, with easy access to the water, and protected. We find a place on the inside bend of the river, where the land falls gently away to a grassy terrace at the end of a rough peninsula. Misty rain mutes the remains of the day. We pitch our tents in the light of our head torches, moving quickly.

Steve proposes a fire.

'Great idea,' I say, but I'm privately sceptical. Everything is damp. We try leaves and kindling and twigs. Nothing takes. I tear a handful of pages from the back of my journal. The pages burn well enough, but nothing else will catch and I'm content to settle into an early, tent-bound evening. Steve, however, is an inspired fire-maker. He finds a large forked branch and sets it horizontally on two rocks he rolls into place to keep it off the ground. In the fork he layers leaves and dead grass, bark and twigs. Beneath the fork he sets up his gas cooker and uses the flame to light the grass and dry the rest of the plant matter, till it too takes. He manoeuvres the little fire below the forked branch and lays larger timber on top for it, in turn, to dry. We prepare our dehydrated meals, take a capful of port each from a flask and barely notice that it's raining.

~

This used to be kangaroo and emu country. James' words won't leave me as I lie in my tent with my journal. This *was* kangaroo and emu country. For millions of years, it was.

Kangaroo: from the Guugu Yimithirr word 'gangurru', recorded by Joseph Banks in his diary as 'kanguru' on 12 July 1770, while the HMS *Endeavour* was beached on the banks of the river the Guugu Yimithirr called Wabalumbaal, in the country James Cook would soon name Cape York. The etymology of *emu* is obscure, perhaps derived from the Portuguese word for ostrich or cassowary, used by early European explorers of Australia, which in turn might have come from the Arabic for 'large bird'. Its scientific name, *Dromaius novaehollandiae*, given to it by the ornithologist who worked on Arthur Phillip's published account of his voyage to Botany Bay, means 'fast-footed New Hollander'.

This used to be kangaroo and emu country. The gravity of what we have done to our fellow creatures should crush us. King Island in Bass Strait was once emu country. They became extinct there in 1805. Kangaroo Island too, where emus became extinct in 1827. Tasmania was also emu country. Until 1865. On an island – even one the size of Tasmania – there's no hiding.

This used to be kangaroo and emu country. Perhaps it still is. But the truth is we saw none today. Not counting a lone wallaby rib.

Day 2

Upper Valley

Distance: 17.27 km

Evening Camp: Place of the thick callistemon

I try to isolate the sounds beyond my tent. Riffling water. Magpie song. Smaller bush birds whose twitching calls I don't recognise. Again the running water, some change in its pitch – a flush of fallen rainwater, perhaps, entering from a small upriver stream. I strain to hear whether Steve is awake, but when I crawl from my tent, I see it's just me up. I find a rise of river pebbles and sit and lace my boots. There's a cool breeze. The clouds part momentarily, flashes of blue between them. The callistemons point the way downstream as unerringly as the running water. Debris in the trees around me marks the flood-line – shoulder height from where I sit – but in this moment the experience of flood is far removed from the delights of this peaceful little spot. Across the stream, a small bush bird with a russet breast and a dark band around its neck pincers a moth in its beak.

I wish I knew what bird it is. And what about the moth? There is a world of moths I know nothing about, family upon family. But what good is knowing the name of the bird and that of the moth

in its beak if you don't understand the relationship between them?
No tableau is populated only by bird and moth. There is season
and climate and land and feast or famine. Every tableau contains
a universe, and perhaps the best you can do is find a river to walk.

~

It's after 9 am before we're packed up and on our way.

The bends in the river, as it comes off the flank of the range,
become wider and longer. We follow as the river sweeps left, presses
against a hard-rock hill and then veers back to the right, before it
reaches a ridge it hasn't yet found a way through and pulls gently
to the left again. The arcs are sometimes so gradual we can barely
discern them. Other times the bends are sharp.

We follow sand, pebble and bottlebrush, thinking no further
ahead than the next footstep. We follow the river until, in front
of us, is a causeway. Yes, of course! Yesterday we'd driven up
this same road, gazing left and right from the causeways as we
rumbled across, engine-and-tyre loud, at driving speed, taking
in fleeting snippets of the width of the river, guessing its depth
through the clear water. Even so, it's a surprise to come across
one again here where the road crosses the river. Now everything
is inverted. Yesterday we were cocooned inside our cars, looking
out as we crossed this ford. Today we're on the river, approaching
it at walking speed, and it is the world of cars that is foreign.

We hesitate as we draw near the road, but there is no choice: to
follow the river, we have to cross the causeway. We step up off the
bank and I feel suddenly exposed up on the concrete, away from the
protective fringe of ti-trees and blue gums. There's no cover at all,
and the sound of our boots slapping on the hard, uniform concrete
is completely alien to the environment we've walked out of, and will,
on the other side of the carriageway, return to. It's like a no-man's-
land, or a wartime transport corridor, though in truth there's very

little traffic – it's a local roadway for local landowners, of whom there are dwindling few this far up the valley.

That was the first causeway – but the river crosses the Western Branch Road thirty-eight times. (Just yesterday, I would have said it was the road crossing the river.) Now, as riverwalkers, these causeways are obstacles forced upon us, slabs of concrete blocking our path. Each has a sign naming it: 'Crossing No 38' or 37, or 36. And each time we have no choice but to step up off the riverbank onto the road.

But the choice we *do* have to make is whether to pass over the causeway and stay on the same side of the river, or to make use of the concrete ford to cross from one side of the river to the other. At the first few crossings, we remain on the same side of the river, out of loyalty to the river and to the journey. The causeways are irritants, and we try to ignore them as best we can.

But it's hard to ignore the number each causeway has been assigned. Hard, too, for Steve and me not to wonder aloud as we approach each one what its number in descending order will be.

I console myself that the urge to number and categorise to help understand is strong. But is this reflexive organising of the world we're passing through unhelpful? Does our too-ready submission to the arbitrary system of identification by which these causeways have become known in signs and maps and roadside conversations, prevent us from other ways of seeing? If, rather than 'Crossing No 35', this place was known for a Dreaming event, or that mound of blue pebbles over there, or that riffle, or the way moonlight in autumn spears off the surface of the deepest part of that pool, might we better understand this place? Might we know the river's stories better – whether sacred or profane – and better place ourselves within those stories? And another question, near the centre of these rippling wonderings: might we in the tradition of Western thought, we great scientific categorisers, be particularly prone to forgetting our stories of place and being? Once, not that very long ago, we existed within a mythological cosmos, undifferentiated from creation and its one-ness, until slowly, as we pursued new forms of organising and knowing, we began to lose touch with our myths, even – for good or ill – to stand outside them.

Among the ancient Greeks, Thales of Miletus was the first to lead us down that path. The world is better explained by science than myth, he thought. And as it happened, water was central to Thales' philosophy – so central that he saw it (in Aristotle's words) as the 'originating principle' of nature. According to Thales, the world and everything in it was composed of a single substance, water; he believed that everything came from water and was finally reduced to water. The importance of water to Thales' thinking may not be such a surprise. His namesake city, ancient Miletus – once in Ionian Greece, now southwest Türkiye – overlooked a famed river, the Maeander, a river so winding, so sinuous, that language itself connects each meandering river back to Thales' home stream.

Then, in the fifth century BCE, Empedocles became the first of the Greek philosophers to conceive of the natural world as composed of four elements: earth, fire, air and water. It was profound – a system that accommodated change and offered a world in elegant, cyclical balance between those four elements – and it was revolutionary, influencing philosophy, medicine and religion. Its echoes resound still: the twelve signs of the zodiac are organised and classified by that same taxonomy. Indeed someone with a closer affinity to that cosmology might ask how I, a Cancerian, one of the four water signs, could *not* feel the pull of the river.

~

We follow the left bank. There are tracks in the soil. Cattle, deer, pigs. We spot the imprints of horseshoes. How fresh are they? Older than last night's rain, but beyond that we can't be more precise. This far up the valley, the country is rugged enough that cattle are still mustered on horseback, rather than motorbike. 'She's pretty hilly and pretty rocky,' a riverwoman from a nearby property, Mae, had told me. 'If there was a good sale for rocks, we'd be rich.'

The river bends to the right. Behind us a crow calls out. I turn.

What is it, friend?

Ark, ark, it answers.

The stony flat on the left of the stream narrows and disappears. We are forced out of the bed and up onto the bank. We pick our way through thickening grass. The bank steepens, becomes cliff, but not sheer. We judge it passable and lean in, our backpacks growing heavier, picking our footholds carefully. I reach for the cliff-face with my left hand to steady myself. The river disappears behind a screen of she-oaks, and then, as we step slowly along a ledge, reappears surprisingly far below. I pause to catch my breath and shake out my legs. I look down across the water at the flats on the inside bend, where the passage would be much easier.

Ah. It's a revelation, sheepishly won. I'd known how rivers like this work but hadn't understood. I've seen it daily, as the river passes through the heart of the city, because Brisbane was first planted on the inside of a river bend. Opposite, on the outside bend, are the cliffs of Kangaroo Point. Only now, looking out, do I understand. As the river carves its way through the landscape and reaches hard rock, it is forced to change direction. On the outside, where it bends away, the banks are higher and steeper, often fringed with thick, impassable vegetation. It is on the inside that sand and silt and basalt stones and pebbles are deposited. It's over the inside bank that floodwaters pour. It's the inside bank that is lower, flatter. I knew it, but not in any way that mattered. Not in any way I could use. Or had to. Now I get it. The inside is much more likely to offer passage. That's where we want to walk.

And so from now on we adopt a different approach at the causeways. The causeways become our friend. We cross them and tack a path on the inside of the river.

~

The bed becomes stony. We stop and pick through a gravel heap looking for signs of sandstone, but this is still basalt country. The river excavates its way back through 250 million years of history. In my hand, 250 million years. How can the human mind possibly reckon with that? I look at the creases on my palm and see contours, gullies, gorges, tributaries. My skin is weathering, and this rounded river pebble is suddenly very heavy.

We bend beneath the weight of our packs and an hour of walking along the pebbly bed, our boots crunching and slipping and turning with each footfall. Though the stones are ancient, the ground is not solid. Steve begins to drift behind me, while ahead the river's long leftwards arc seems without end. And then, above the bank to our left, human voices, men.

I stop abruptly and motion Steve to do the same. We hear the sound of a vehicle door slamming shut. More muffled talk. I listen but can't make out words. Yet the voices ahead are unsettling, the words breaking up on the breeze before they reach me. I don't know whether these men are farmers or hunters, and what the sight of our heads rising over the bank might mean to them. But there's another thing. Already – after only a day and a half – my sensibilities are shifting. Sounds as mundane as these – raised voices and car doors closing – are intruders from a different world. Steve joins me, and we sit. We have time to bide. There is banging, things being thrown into a truck's tray. Engines start, one, two. I ready myself in case the vehicles come this way. But they don't. They reverse, turn, change gears, move away, and soon enough these hunters or roadworkers are gone and all that remains of their having been here, just downstream, is a campsite of compressed grass and a campfire carefully doused with water.

~

The stony pan widens, the stream and a fringe of stooped callistemon on our left. We settle into a rhythm, talk superfluous. Or rather, perhaps it's the sound of our bootfall that is speaking, each to each. A seven-foot-long red-bellied black snake crosses our path just ahead, moving quickly over the pebble bed before disappearing into the long grass to our right. From the grass a pheasant coucal takes flight. We barely break stride, mammals and reptile and bird making space for each other.

By midafternoon we're flagging. A grey cloud descends upon us. The stony bed shifts beneath our feet with each step. It's tiring. The river curves right, widens, deepens. We're on the inside of the bend, where the flood has dumped the river stones into heaps. On the far bank, a cliff begins to rise from the water. We slough off our backpacks, slide out of our boots and hang our clothes over a flood-battered log lying askew on a pile of pebbles.

There's nothing more natural, or more joyful in its simplicity, than swimming in a river.

But I hear the faint echoes of warnings. In the city – over three hundred kilometres downstream – swimming in the river is unsafe, unwise, forbidden. Each time a tourist or inebriated student is lost in the river it is not merely a tragedy but becomes a sober caution: the river is polluted, its currents are unpredictable, you're invisible to watercraft, the bull sharks will get you.

But here the water is clear and cool and deep and we are hot and tired and our feet are sore. So we swim. I duck-dive and hold my breath and wait for the air in my chest to sharpen, the blood to pound. I rise and burst and splash and suck and shake my head. We laugh. We stroke, reaching out again and again. We touch the face of the cliff on the other side, turn back, and when we pull ourselves out of the river, we are new.

~

There's an hour's walking left in the day. The river flows beside us. It bends. Flows beneath a causeway. Bends some more.

But a river doesn't bend only when it reaches hard rock and is forced to turn away. Sometimes a river winds because it wants to. Some rivers just want to meander for a bit. Swerving, snaking, curving. Others are relatively straight. The short rivers that plummet off the ranges of New Zealand are straight. The Brisbane – like most Australian rivers – is sinuous. Why?

As inheritors of the tradition of Thales of Miletus, we men and women of science study rivers. That great fluvial hydrologist and sometime painter, Leonardo da Vinci, filled his notebooks with drawings and musings about the flow of water. A meandering river looks over the shoulder of the *Mona Lisa*. Now, we create disciplines and university courses and even academic institutes to study rivers. There is no end to our quest for knowledge.

We categorise rivers and interrogate their workings. We've created numerical models to try to understand and predict their flow. One model looks at the relationship between the roughness of the terrain and the angle of the slope a river traverses to determine its sinuosity. Another factors in the diameter of the sedimentary particles transported by the river and asserts a law: that there is a regular downstream decrease in the size of the sedimentary particles which is related to the decrease in the slope of the stream. Or, more simply, the finer the sediment, and the flatter the slope – conditions that occur more often in the lower reaches of a river – the more likely a river is to meander lazily this way and that. We've even devised indices to determine how sinuous a river is, one of the neatest being to divide its entire twisting and turning length by the straight-line length of its river valley. The Brisbane is very sinuous indeed.

But look carefully at water when it moves round a bend: what is that mesmerising corkscrew movement, that secondary, helical current? The one da Vinci sketched again and again, and which momentarily captured Einstein's attention in 1926, when – after studying the movement of tea-leaves in a stirred teacup and noticing how they settled in the centre rather than being pushed to the outside as one might expect – he wrote a paper that explained how helical flows develop in a meandering river. The swifter portions of a stream are driven to the outside bend, and that's why erosion is greater there. This helical current is one that hydrologists and applied mathematicians have swum in for a century or more. Theories, formulae, ratios, research, testing. Pebbles of knowledge gathering on a great river's beach.

~

Steve and I walk along the beach until, eventually, we break for camp. We take off our boots, feel the sand and river stones on the soles of our feet, and are grateful to be here.

Do campsites, I wonder, like rivers, operate according to some universal law, adapted to local conditions and personal preference? The routine and methods are the same the world over. Choosing your site. Sliding out of your backpack for the day, that relief. Then pitching your tents.

As for us, Steve and I both have light, one-person shelters, mine grey, his bright orange. Once we've erected our tents, we lay out our inflatable sleeping mats and our sleeping bags, a silk sleeping sheet inside each bag. Steve stows his pack in his tent, at the foot; I put mine outside, under a small awning. Inside our packs are our changes of clothes and all our camping bits and pieces: torch, matches, toiletries, first-aid kit, micro-towel, GPS, pocketknife and, for me, notebooks. Each of us has enough food for six days – the five days we'll be walking until Linville, and an extra day's worth 'just in case' – and the utensils we need to cook it. How to keep gear organised is the stuff of preference. For both of us, it's a combination of bags, plastic ziplock sleeves and compartments in our packs themselves. We swap our boots for light sandals, easier to wear while eating and talking and staring into campfires.

Tonight we set up camp between the stream and a thick band of callistemon. I slide my three-litre water bladder from the frame of the backpack where it's stored during the day – a valved tube slung over my shoulder, easily accessible as I walk – and go down to the stream to refill it.

After water, fire. I'm using a brand of gas burner with a built-in pot that's evolved for lightness and hyper-efficiency. Cooking a meal is as simple as taking a minute to bring half-a-litre of water to the boil on my Jetboil, pouring the water into my dehydrated meal pack, stirring and sealing it for ten minutes to rehydrate. And that's it. My meal's packet doubles as plate.

But a campfire is a different thing altogether. Though we may not need one for cooking, we need it for warmth – it's that or crawl into our sleeping bags for an early night – and to keep insects and animals away while we talk. And, perhaps most importantly, to look into. To become mesmerised, to be transformed.

We gather fuel for the fire. Steve tells me about a course he did many years ago in San Francisco at which he'd been taught a fire-making technique used by North American First Nations and settler peoples. He produces a little pack of material – a flint stone, a piece of steel, char cloth and tinder. He strikes the flint stone with the steel and catches a spark in the char cloth. The flint and striker fall to the ground, his complete attention on the spark. He blows the spark into an ember, transfers the ember to the tinder, still blowing gently, at just the right angle, till it grows into a flame, then carries the flame to the fuel we've gathered. He makes it look easy.

'The hardest part,' Steve says, laughing heartily, 'is remembering where you've dropped your flint and striker so you don't lose them.'

We look into the fire together, seeing, not seeing, unseeing.

Day 3

Upper Valley

Distance: 13.16 km

Evening Camp: Place of the island in the stream

In the mornings, a campsite operates in reverse. Perhaps a fire. Breakfast. Coffee or tea. Cleaning teeth. Putting on the day's hiking clothes. The lacing of boots and clipping of gaiters. Packing up – tents and flies last to give them the best chance of drying. I savour all this. The joy of having so few possessions, and the knowledge that each object in my pack has a purpose. The care that each thing deserves. The slowness that care requires, the deliberateness of it. And how, when packing up a campsite, there is always a bird on a branch somewhere, watching.

~

We're becoming more familiar with our gear and the routines of camp and are ready to leave a little earlier than yesterday. Again I have to pack my tent wet – from the evening rain, from the morning dew. We step back through the stream to the left bank and continue our journey. The watercourse presses against a hill covered in flowering red Natal grass. It's not a native grass, I know that, but lit by the morning sun, glowing, it is beautiful.

Steve walks ahead of me as I pause to dictate into my phone's recorder something I don't want to forget. He's in shorts, a navy collared shirt, a soft hat. His towel is draped over the back of his pack to dry. He's spent hundreds of days in the bush across his life, and I suspect he's as comfortable out here, now, as anywhere on the planet.

We come to the first barbed wire fence of the day. We slide out of our packs and roll them under. I part the bottom two strands of wire for Steve, my left boot pressing the lower wire down, left arm raising the higher strand up. Steve bends and steps through and, once on the other side, creates a gap for me to follow him.

Already we're adept in the fine judgements we have to make in helping each other navigate these obstacles – first gauging how taut the wires are, and thus whether to step over, bend through or crawl under the fence, and then whether we'll have to take our packs off to throw them over the fence or slide them beneath the lowest strand. The objective of each fence crossing is to expend the minimum energy possible. Having to get out of your pack, throw it over and then shimmy on your stomach under the bottom wire before getting up and shouldering your pack again takes the most out of you. Stepping over a loosely wired fence with pack still on your back is best. But just how low the top strand of barbed wire needs to be to step over it differs from person to person. In this, the god of barbed wire fences favours the long-legged. The consequence of failure – a barb catching the crotch of your trousers, a leg planted either side of the fence – is serious enough that the decision to attempt the manoeuvre can't be delegated. But with each successive fence, these judgements become increasingly instinctive.

You learn to trust someone after helping each other over a hundred barbed wire fences.

~

The river bevels downstream through its basalt bed. We stop for morning tea at a mound of pebbles on an elbow of the river, unclip our packs, sit down and take out muesli bars. We bend to the stones and begin turning them over, selecting a dozen or so that fit neatly into the palm. Steve takes out his steel striker, and we test our specimen pebbles. Only one sparks. Not as prolifically as Steve's North American flint stone, but we're hopeful.

We reach another causeway, crossing number 17. We're counting them down now. On its downstream side is a cascade of concrete steps. Usually, like now, the river's flow passes through the huge pipe running beneath the causeway. However, when it starts to flood, the water overtops it before falling down these tiered steps, losing energy that might otherwise erode the bed and scour the banks below the ford. Smart, I think. But then I remember that if the road hadn't been laid along the length of the valley in the first place, there wouldn't be a problem that needed mitigation. There wouldn't be need for an engineering solution to an engineering problem.

~

The beats of boot and blood fall into synch.

Beside us the river burbles and hums and trickles and runs.

There are spiritual traditions that believe some words resonate with a frequency which, if spoken or chanted or prayed aloud, open a doorway to another consciousness. Words for the name of God. Certain ancient chants, *Om mani padme hum* and *Hail Mary full of grace* and calls to prayer. Lines of poems so perfectly shaped that, once heard, cannot be forgotten. The names of some rivers are, I'm sure, like that: rivers whose names are ancient, rivers named as language itself emerged and first began to flow from human tongues, language born, perhaps, from a generative universe of burble and flow and watersong. Rhine (Gaulish Rēnos, from the Proto-Celtic rei, that which flows) and Volga (Proto-Slavic, wetness, moisture) and Tigris

(Sumerian, running water, swift water) and Indus (Sanskrit, *síndhu*, river) and Barka (Barkindji, also, simply, river).

There are rivers named for their size: the Murrumbidgee (big water in Wiradjuri), the Mississippi (the French rendering of great river, *misi-ziibi*, in the Ojibwe language) and the Niger (*gher n gheren* in Tuareg, river of rivers, shortened to *ngher*). Or their colour: the Thames (from the Brittonic Tamesas, stemming from a word for dark), the Moskva (black river, as one theory has it, in the tongue of the river's earliest people), China's Yellow River, Vietnam's Red River, or more recently the Colorado (literally coloured, in Spanish, that colour also being red).

I muse. Perhaps each language of the world was born of river, and the differences in each language and dialect reflect each river's uniqueness, the different sounds of water flowing through unique geographies. I think back to Europe, where it seems there is a country for every language, and a language for every major river's catchment. The Thames and English, the Seine and French, the Rhine and German, the Vistula and Polish. Exceptions, like the Danube, seem to prove the rule. So many great rivers that never meet, in catchments bounded by ranges and shorelines. So a catchment becomes a community becomes a people becomes a culture; the river is that people's great trade route, the site of its cities, its capitals. And in my musing the river is the ancient source-song of language itself.

~

Up ahead a wader stands at the water's edge with its long, curved beak, its white breast and leggings aglare in the sun. We cross the river to find a spot for morning tea. Before us, grazing on the bank, are a dozen red Angus cows. We move slowly. We don't want to spook them. What we don't see at first is the property boundary a little further on, a barbed wire fence, trapping them. They're becoming agitated, so we stop completely, then watch them leap into the river. They swim in single file, one after another.

On the other side of a gully with a name, Charlie's Gully (who was Charlie?), a couple of large prickly pear trees stand on a low hill. Steve slices a piece of bright red fruit off one of their pads and opens it with his pocketknife. Delicious. Nearby, a dozen dark birds explore the leaf litter on the edge of a stand of river-fringe blue gums.

Apostlebirds, I think.

'White-winged choughs,' Steve says, and he'll be right. From this distance the birds ignore us. Red-eyed, curve-beaked, black feathered. Yes, apostlebirds are lighter in colour than these, grey, with shorter beaks. A lifetime ago my cousins and I used to chase apostlebirds – which live in groups of about twelve, like Christ's chosen disciples – shooting at them with crude homemade arrows, invariably missing. Back then we knew them as 'happy jacks'. I suspect our Catholic consciences wouldn't have allowed us to shoot at them if we'd known their alternative name. Steve and I begin moving again. The birds rise from the ground, their white under-wings suddenly visible.

Just before midday we reach the point where the Eastern Branch joins the Western Branch of the river. We hadn't realised it was this close. It's a surprise. And a problem. A couple of kilometres back we'd decided to cross over from the right side of our branch of the river to the left, which meant the river was on our right. Now that we're at the confluence, with the Eastern Branch coming in at a ninety-degree angle from our left, we're stuck on a grassy promontory between the two branches, water either side of us. We survey them both. The Eastern Branch is as wide, perhaps a little wider, than the Western, and both are too deep to wade across. Bugger. We resort to checking my satellite map. It looks like there are cattle yards half-a-kilometre up the Eastern Branch, and perhaps there's even a crossing nearby. So we head upstream and find a ford across the river near the yards. The river is running over it, but it's shallow enough for us to wade through. So we cross the Eastern Branch and turn back downstream, back to the confluence, where a grebe hugs a shaded bank and a buzz of insects hovers over the water.

Downriver is a broad flood plain, strewn with basalt rocks. We take a long causeway back to the south bank, and find two horses with white blazes on their foreheads. They turn their heads, curious. Who are we? The low cloud of the last two days breaks and gives way to direct sunlight and blue sky. We crunch our way along the stony plain, the sun on our head and shoulders, the heat reverberating off the rocks, conditions significantly hotter down here in the bed than up on the grassy bank.

We spy a deer and leave the bed and find a fig tree and stop for lunch.

The deer is standing calmly in the middle of a mixed herd of cattle, Angus and Hereford. But why? Are there other deer nearby? Has this one paused as it passed through, the grazing here worth lingering for? As we reach into our packs for our provisions, the deer, and the cattle, slowly move away.

Steve cuts chunks of cheddar onto biscuits. I pry a couple of triangles of soft white cheese from their cardboard wheel. After unpeeling the foil wrapping, I spread the cheese over my pita bread. I'd discovered this brand thirty years ago in Algeria, where I'd first smeared it on flatbread while sitting in dry oueds in the Sahara – a processed remnant of French colonialism, but one I've been fond of ever since: *La Vache Qui Rit*. The laughing cow.

After lunch, we see the cattle again on the other side of the river. The lone deer is still among them, at the tail of the group as they pick at the grass, raise their heads to inspect us, and then move on, untroubled. The deer is at ease, one of the herd.

How does this happen? Is there any limit to interspecies relationships? There are the obvious relationships: predator and prey, host and parasite, competitors – for food and habitat. But cooperators too, in the grand narrative of creation – over the hill, an egret perches on a Brahman's back, picking ticks off its hide, to their mutual benefit. But what connects deer and cattle here? Perhaps the deer's mother was shot by hunters when it was young, and a cow in the herd who'd lost a calf had some maternal attention to give a motherless fawn? An interspecies adoption? Or maybe it is just companionship the deer seeks, the cattle unperturbed by its presence. I wonder, I guess. Boundaries dissolve. That which appears fixed, changes.

~

While this may be kangaroo, emu and cattle country, I can't deny the thrill of spotting a deer for the first time.

I know it's an introduced species. That there are thousands of them up and down the river valley, as well as in the Mary River valley immediately to the north, descendants of six ancestors – two stags and four hinds – introduced (or 'liberated', to use the quaint phrase of the day) in 1873. Or so the legend goes. It may mostly be true. The story of their arrival is charming, or dispiriting, or devastating. Or all three.

It is charming in its gesture to the importance of personal relationships; nice that a woman, Victoria, albeit a queen, might be so touched to have a colony – Queensland – named after her that she would select deer from her private estate at Windsor Great Park as a gift to the colony, bestowing it on one of the first colonial families to have settled in the valley. See, today, the deer on the state's official coat of arms? Origin stories matter. Good ones – true ones – have the power to gather and amplify and protect and preserve and celebrate. The very best are sacramental.

It is dispiriting, however, that the tale of the deer's coming should be so narrow in its compass; that the gift was confined to a few wealthy pastoralists, for 'additional food and sport'.

It is also, with the passage of the years, devastating. Red deer (*Cervus elaphus*) were introduced to colonies far and wide: Australia and New Zealand and Argentina and Chile. Far and wide they have multiplied and become feral. They trample seedlings and ringbark saplings and out-compete native mammals for food. They eat out plants and spread weeds and despoil water sources. Of course they need to be managed, which probably means culled, which means – one way or another – hunted.

It's a challenge to hold competing experiences like this in tension: the thrill of seeing a deer for the first time, the grotesquery of its undemocratic introduction to this place, and the knowledge of the pressures these thousands of hooved ruminants now place upon this ecosystem. My ecosystem.

~

We're only a kilometre or two further downriver, deer on our minds, when Steve spots a pair of antlers on the opposite bank, two pale bony points, about fifteen centimetres long, protruding from a small clearing of low grass a metre or so back from the waterline. Deer grow a fresh set of antlers each year, dropping their old rack – their 'cast-offs' – in spring, between August and November up the valley.

We look at each other. We understand one another well enough now to know we're asking ourselves the same question, both gathering and weighing information. The river's too deep and wide to cross here. The last possible fording was a couple of hundred metres back. Those metres have been tough, wending slowly through a thick swathe of callistemon. It's taken precious energy. Backtracking takes a different type of energy, and the sun is continuing to move its way higher through the sky. The journey to those antlers and then back to where we're now standing would mean treading the same path three times. But cast-offs! Our first.

And having navigated the route through the callistemon forest once, it's now familiar. So we turn.

When we reach the clearing, it takes a while to find the antlers again, a lesson in how some things are harder to find the closer you get. Eventually we locate them. I reach for the rightmost antler to lift it, but it won't come. I kneel and clear away the grass and soon realise that what we'd first spotted across the river were not cast-offs at all. Hidden in the grass, half-buried, is the stag's skull. What we'd taken for the antlers' tips were actually just its brow tines – the tines closest to the coronets on the skull from which the antlers grow. My blood quickens as I dig, my hands scratching and scrambling until they pull, finally, a complete set of antlers from the ground, almost a metre in length, still attached to the deer's skull.

So, I think, a victim of February's flood. I imagine this creature caught in a wild torrent, unable to swim free, drowned. Then buried in the sediment laid down by the flooding.

Before we leave Steve carefully places the stag's skull and antlers at the foot of a large gum.

~

By midafternoon we're tiring. Steve says he's not thinking clearly anymore. Exhaustion does that to all of us, and Steve knows himself well. We take off our boots and wade across the river, the stones sharp under our soft, tenderised feet. We think we're crossing to a promising camping spot on the right bank but discover we're on an island, the river flowing either side of us. It's not just promising – it's perfect. We drape our tents over callistemon branches to dry and lie back, close our eyes and listen to the water running by.

~

That evening the river stone we've carried with us so hopefully won't spark at all. We both try, both fail. Instead, Steve starts the campfire with his lighter. With the fire going, we prepare our meals: dehydrated rice and bolognaise sauce for Steve, creamy pasta for me.

The night sky clears, then deepens. After dinner, Steve produces his port and we each take a nip.

We step away from the fire to look for the Dark Emu in the sky. It's a brilliant sky, free of the city's ambient light, which too often blinds rather than illuminates. I look up. After a lifetime of seeing celestial shapes composed of stars, of focusing on the stars themselves and the relationships between them, it's difficult at first to see the space between them and to find the emu in the darkness. To focus not on presence, but on absence. But suddenly you see it, the emu's head – the dark patch that is the Coalsack Nebula – is there to the left of the Southern Cross, and there are its beak and long neck stretching through the two Pointers, and then its breast and its body and its legs all appear, and once you see it, you never unsee it. This used to be kangaroo and emu country. Perhaps it always will be.

~

We linger in the night sky, pointing out satellites to each other, artificial stars moving steadily in their orbit, their reflected sunlight. There are thousands of them up there now. Each year adds hundreds more. Alone as we are, here on this tiny island in the middle of the river, a satellite moving across the night sky is a reminder that nowhere is remote any longer and, short of apocalypse, never again will be. Remoteness, wilderness, virgin spaces – all now the stuff of myth. Or, as I reflect while lying in my tent, perhaps they always have been – the myths of city dwellers and Western rationalists and colonists and humans bent on seeing ourselves as special and separate from the rest of Creation.

Day 4

Into Avoca Vale

Distance: 14.55 km

Evening Camp: Place where mist skirts the hills

The river calls. Or the walk does. Or maybe I've just woken this morning with an urge to get moving. To *be* moving, to not be still.

For the first time we see, punctuating the grassy verge, large patches of turned earth – five or six metres long, some of them, and a couple of metres wide. They've been worked as thoroughly as if they'd been ploughed – the grass ripped up, stones worried out of the ground, the soil pocked and divoted. Pigs.

There is evidence of pigs for a couple of hundred metres, all the way down to where Cooyar Creek joins the river from the southwest. Cooyar is the creek some say is the true source of the river. It rises in the distant Bunya Mountains, once the site of three-yearly bunya nut festivals at which people from different Aboriginal Nations would gather for trade and celebration. It carries more water than either the Western or the Eastern branches. I wonder, standing here now, whether this junction might be a waypoint of an ancient travel route. But pigs interrupt my wondering. Because here, now, we see them: a dozen ferals rooting in the sloping bank on the other side of the river. When they notice us they rumble away, their black-and-white coats disappearing behind a thicket of lantana.

~

It's been three-and-a-half days and we haven't yet seen another human, but all around the eyes of birds and other animals fix on us, and we, sometimes, see them. There's a trail in the sand, clean-edged and fresh, not yet blurred by wind or rain or the passage of other creatures. Snake, I think, and follow where the track leads. But no, it's a goanna, motionless. I too stop and raise my hand to signal to Steve. That's enough to set the goanna running, head and neck and chest held high in that unique way of his.

We step onto a great bed of blue and grey river stones. They're smooth and hard beneath our feet but constantly move under our weight, slipping and sliding and rolling. They're remarkably uniform – about the size of my fist, some a little smaller, a few larger. I pick up a stone, cool in my palm, grey as the heavens. Ten thousand of them on this bed of the river. How many eons of time do they bear? How much wind and rain and sun has scored them? How much coursing river water has run over them? Smoothing, buffing, polishing. And how far, I wonder, have they been carried? Flood after great flood, rolling them just a little further downstream. I bend again, pick another up, carry it in the palm of my infinitesimally young hand. How perfectly it seems to fit.

~

I'm beginning to understand that I'm walking not beside the river, as I'd first conceived this journey, but *in* it. I've been walking down in the bed of the river for much of the last few days. Yes, it's been dry – sandy or gravel or grassed – but it'll be covered with water next time it floods. *Flood.* Suddenly, down here in the river, the whole notion of flooding feels wrong. Floods aren't exceptional events, aberrations from the normal course of the seasons. Flooding is itself a season, one of the seasons of the river, completely natural, completely to be expected, completely necessary.

~

A plaque on the bridge reads: *This bridge stands as a memorial to the men who in the 1940s and 1950s constructed these roads and causeways from Linville to the head of the river to access pine timber for the Dept of Forestry. This plaque acknowledges their efforts and achievements.*

Just downstream of the bridge we read another sign, this one torn off its bridge-side post in February and washed halfway across a paddock, its upturned face stained with intricate swirls of mud: *ROAD SUBJECT TO FLOODING. INDICATORS SHOW DEPTH.*

Evidence of the flood is everywhere. There is the manufactured human debris like the road sign and the fencing ripped from the earth and the coloured pieces of plastic littering the way like confetti. But it's the chronicle of the flood on the landscape that catches our breath, again and again. There's the high-water mark on the trees and bushes where thatches of dried grass have been caught, and the scars on tree trunks where large objects have gouged out flesh on their brutal downriver path. By the water a generation of callistemons points floodways. Blue gums stand unsteadily, the earth eroded where the water has eddied and ripped out the soil around their bases.

And then, high on the left bank, we come across a gate at the end of a boundary fence. A large portion of the bank has fallen away and taken with it the corner post nearest the river, leaving half the gate to swing out over the river, ten metres above the water. Nothing to fence in or let pass but air. Dried grass has been caught on the gate's upstream side, so much of it that the debris completely covers the diamond-patterned wiring and gives the impression of a gate constructed entirely of grass. We push the grass gate open. It creaks on hidden hinges. We manoeuvre our way past, holding onto the sole remaining gatepost with both hands, our chests pressing close against it as we swivel round, our packs momentarily suspended above the river.

~

Down in the bed, we begin talking about the city up ahead, the city at the end of the river, the city where we both live and where Steve is a professor at the sandstone university nestled in one of the river's crooks. Steve loves rivers. He studies them, gathering data about wastewater and stormwater and stored water and the energy required to store and transport and use and recycle that water. He processes that data, analyses it, lives with it. Thinks about the way a river city like Brisbane functions and whether its water supplies are secure, whether its use of water is sustainable. But he also canoes them. And so our conversation meanders, at riverwalking pace. We talk about how rivers are the principal wildlife corridors through a city; how traditional walking routes followed them; how on da Vinci's maps of waterways he compared them to trees; how Brisbane, with its subtropical storm patterns and variable rain, is such a flood-prone capital. We talk about natural water cycles and anthropogenic ones and how the study of paleoclimates reveals that this corner of the world has had deeper and longer droughts than we care to accept. That years of dry and thirst will return. And that desalination is too expensive an answer to preparing a city like Brisbane for drought.

'The city is growing so quickly, using more water each year. That puts pressure on our river, enormous pressure. As our population grows and our need for water increases, those pressures are unsustainable. But there's a paradox. Each year more rain falls on Brisbane than the city needs. About four times as much. And most of that rain just washes away down the stormwater system. We need to collect more water – more of the water that falls on our roofs and on our streets. Before about 1910, every house in the city had a rainwater tank. But in 1910, the city also had dengue-fever outbreaks. Too many of those rusting metal tanks back then bred too many mosquitoes.'

Steve talks while he's walking, with a calmness shot through with enthusiasm.

'Of course we need to reduce the pressure we're putting on our rivers. Really, we have no choice,' he says. 'But we also want to make our cities more liveable and more beautiful. We can do both.'

~

On the satellite map there's a river property, Rathburnie, just ahead marked *Nature Reserve*. A green oak tree icon on the app's map emphasises both its ecological status and, annoyingly, a persisting imperial reach. As if a genus of tree as foreign to this land as the oak can signify anything useful. The property had once been owned by a couple who between them were its stewards for three-quarters of a century. Cattle farmers and environmental advocates, they'd wanted the property to 'inspire landholders to rethink land management on this ancient continent'. Now a government website describes it as a wetland habitat, and lists 167 species of plants, animals and fungii that can be found on the seven square kilometres of the reserve. Among the mammals are whiptail wallabies and Herbert's rock-wallabies, squirrel gliders, platypus and koalas. Steve has spent four-and-a-half days forlornly scanning forks of blue gums for koalas, so he's keen to visit on our way through.

I had asked a local farmer about it last year.

'A wildlife refuge?' he'd scoffed. After the last of the owners died, he told me, the World Wildlife Fund (WWF) promptly leased it so it could continue to be run as an agricultural property, raising cattle and lucerne. Late last year the WWF offloaded it completely.

'So it's not a wildlife reserve?' I couldn't quite believe it.

The farmer smiled patiently. 'It's a cattle and lucerne farm.'

I don't know if the farmer was a realist or a cynic or privately antagonistic to ecological land management. But the previous owners' vision was that it could be both farm and reserve, and there's a conservation covenant on the property – theoretically forever – and I want to believe.

We follow the river into the reserve, the sun falling. We breathe it in. There's no doubting – this is a very beautiful property. We crane out necks looking for koalas, but if they're up there in the eucalypts this afternoon, we are destined not to see them.

~

Duckadang perches on the left bank, high. We say it aloud as we come off the river, repeating it back to each other, this strange, beautifully weighted word. Duckadang is a camp – a collection of dormitory-style halls and a covered communal eating area – run by a community services club. The land had been donated to the club in the 1970s by the owners of the reserve we've just come through. It's a bed and shower for us tonight, and a chance to wash clothes. Duckadang was, I learn from a plaque, 'named after an Aboriginal Tribal Elder best remembered for his persistence, determination and sense of duty'. I can learn nothing more of its name, or the man, and find myself wondering about this story. It feels a little too earnest. Duty to whom, I want to know, and to what?

In the evening, Ali, a local landowner I'd spoken with while planning the walk, visits us at Duckadang on her way home from the nearby town where she works as a teacher's aide. She's just seen the carcass of a stag on the side of the road. Andrew, the camp's caretaker, saw it too – it's impossible to miss, he says. Hunters had dragged the creature up to the side of the river road and dumped it there after severing its head and antlers for a trophy. Dumped as thoughtlessly as tossing litter from a car. Dumped it where it will rot and create a stench so great the locals will have to wind their windows up as they drive past. Dumped it without heed to all that is sacred in life and in death. Ali and Andrew are angry. But there's more to it than that. Their spirits have been wounded too.

I lie in my lower bunk bed in the dormitory and stare at the springs of the bed above me and feel something of Ali's and Andrew's pained fury. It wasn't flood that took yesterday's deer, I think to myself now, but hunters. A different narrative forms. A stag in a rifle sight. A round finding flesh but only wounding, not killing immediately. The stag fleeing, his hunters unable to track him; the stag coming down to the river, dry from death-thirst.

Day 5

Avoca Vale to Linville

Distance: 17.7 km

Evening Camp: Place of the wide-verandahed hotel

Today could be a long day. Ali drops by again on her way to work and tells us she once farewelled her kids from in front of her property as they set out to canoe into Linville. They never made it: it was too far, the water too shallow too often, with too much portaging, and when the search party was sent out they discovered the kids hadn't even made it halfway. So Ali is concerned for us, and perhaps we are the beneficiaries of some lingering maternal guilt, because she offers to take some of our heavier items and drop them at the Linville Hotel, where we've booked in for the night. There's no rule of riverwalking that says one can't accept offers like Ali's. Rather, one of the organising principles of this walk is the opposite – to be open to what the river and the journey bring. With gratitude we unload the heaviest and bulkiest of our gear into the back of Ali's four-wheel drive. Perhaps we'll see each other for dinner at the pub tonight. Or perhaps we won't.

We leave the camp and cross the causeway and are back on the right bank of the river. A cattle pad leads us. On the dewy ground there's a matting of spiderwebs – tiny, delicate. When the track forks, Steve takes the higher path, his eye set for deer. A herd of rusa (*Cervus timorensis*) roams in the foothills near here. Ali permits them onto her property and won't allow hunting. She's creating a refuge.

~

The bed is sandy. Our boot prints join the tracks of a thousand wading birds. What patterns might be discernible in our comings and goings? A mandala, or labyrinth of way-making. Steve introduces me to a black-fronted dotterel. Good morning, little bird. I've seen you so often over the years, and now it's good to know your name.

A dog barks behind us, and the dotterel darts away. When we turn, we see not just one dog trotting towards us, but four – a tan-and-white collie, a terrier, a dachshund and a dashie-collie cross. Soon they're sniffing our feet, and it's clear they're after adventure. What can we do but indulge them? And ourselves: it's good to have their company for a while. We're tracking the river on a long, sweeping bend. The dogs run ahead to explore, wait for us to catch up, then take the lead once again. After a kilometre, we begin to wonder where the end of the dogs' territory might be and when they'll turn around and head home.

The river comes to a concrete causeway.

As we're about to step up out of the bed onto the concrete slab, a LandCruiser approaches from up the river valley. We pause to let it pass. It slows, and I recognise the driver. Bruce is a respected valley riverman. He knows the river as well as anyone – through family history, through living on its banks and through his work for the non-government organisation responsible for the health of waterways in this part of the state. We'd passed through his property a day or so ago; now we greet each other and talk about the season, flood damage, the route. Bruce recognises the dogs – they're Ali's, and they'll make their own way home when they're ready.

Steve asks Bruce about the flow of the river here at the causeway.

'It flows most summers, but not as high as this,' Bruce says. 'In winter, it will usually retreat to a series of ponds, but this year the ground's still saturated from summer and all the rain since. The water's still making its way out of the system.'

Bruce says that, following the February flood, tilapia have begun migrating as high as his place. *Oreochromis mossambicus*, Mozambique tilapia. Once an exotic aquarium fish, it is now a marauder of the waterways. Tilapia reproduce rapidly, are omnivorous and aggressively dominate a river, claiming habitat and food, disturbing plant beds, outcompeting native fish.

'Is that depressing?' I ask.

Bruce doesn't strike me as the sort of bloke who'd readily use that kind of language, and I know immediately it was a lazy choice of word, but he looks at me straight and says 'yes'. And then sighs. They've been expecting it, having seen the fish make its way, bit by bit, up the river.

We continue downstream, the dogs our companions still. Another kilometre, another bend navigated, another causeway up ahead, and the dogs show no sign of returning home. We know what we have to do. It's not easy, but we turn on them. We stand taller and deepen our voices. 'Get home, garn, get home!' They scurry away some, but after we've turned our backs to resume our walk, they come again. We yell and wave them away, and when they're at their furthest we throw stones towards them. We think we've had success, but as we step up onto the causeway, they bound enthusiastically back towards us. While we pause to reconsider our plan, a truck approaches from down the river road. The driver knows these dogs too and gets out of the cab to help push them across the causeway so they can take a short cut home via the road. He climbs back into the truck and crosses the ford, driving the dogs in front of him.

We hurry back down into the riverbed so the dogs can't see us if they're tempted to come back, and we are soon obscured from the road by a stand of she-oaks.

We weave our way through the river-trees. So closely do they stand together that it's impossible to chart a course between them without our shoulders bumping against their trunks. A clearing opens up in front of us. We step out into the cloud-muted sunshine. There, waiting patiently for us in the clearing, is the mottled dachshund cross. Just him. Well, we think, there's little more we can do. So we fashion a new plan: the dog can accompany us to Linville, where we'll tie him up to wait for his owner that evening.

So he joins us. He reminds us of a piglet, and that's what we nickname him.

Piglet gets through the first barbed wire fence before us. We follow, rolling our packs under, then parting the strands for each other to bend through. On the other side we shoulder our packs and are about to turn on our heels when we see, twenty metres back, a red deer hind and her fawn who have been quietly following us.

~

We spot mussel shells on the bed as we walk. The husk of a dragonfly. A saw-shelled turtle's carapace.

I stop and look. *Really* look. See the patterning of its shell for the first time: the three ladders of dark plates running down the length of the carapace and the ring of smaller plates around its hem that resemble the serrated blades of a saw. See the sutures between the plates. Turn it over. See how beautiful the underside of the marginal plates are, dark still but shot with bands of orange. Why, I wonder, why? I notice the two curved bridges between the carapace and the undershell, which I later learn is called the plastron, but the word is ugly and I prefer breastplate, even though it's composed of more than a single plate. This one's breastplate is pale, almost white, with two rows of scutes and spikes either side of its neck hole.

So much for the species, but what about *this* creature? What stories do those three cuts across its pale underplating tell? And what about the chipped rim of its shell at the front left of its carapace? What battles and falls might these scars hint at? And what happened here, where I'm kneeling? Is this the site of a feral pig's mauling? Or is this where kids, playing in the river, dragged the turtle after netting it, only to learn they'd wounded it in the process? Or is this merely the last place an old and frail turtle reached before it could plod no more?

~

There's a story about another chelonian, a tortoise who lived on the banks of the river much further downstream, in the city – a giant Galapagos tortoise collected by Charles Darwin as a pet while he surveyed the natural world on the HMS *Beagle* and began to muse about natural selection. The story goes that in 1831 Darwin befriended the first lieutenant on the *Beagle* during her second voyage, a fellow by the name of John Wickham. Darwin and Wickham shared a cabin. Together they sailed and they sailed. Around the earth they sailed, from 27 December 1831 to 2 October 1836, wondering, querying, surveying, sketching, recording. The following year, Wickham was promoted to captain for the *Beagle's* third voyage, but by then Darwin had disembarked in Plymouth with his pet tortoises to find his land legs, write up his journals and wonder about transmutation. *The Voyage of the Beagle* was published in 1839. Recalling the Galapagos Islands and his tortoises, he wrote:

> ... the most remarkable feature in the natural history of this archipelago ... is that the different islands to a considerable extent

are inhabited by a different set of beings … The inhabitants … state that they can distinguish the tortoises from the different islands, and that they differ not only in size but in other characters. … The specimens that I brought from three islands were young ones, and probably owing to this cause, neither Mr Gray [the British Museum's reptile expert whom he consulted] nor myself could find in them any specific differences … But it is the circumstances, that several of the islands possess their own species of the tortoise, mocking-thrush, finches, and numerous plants, these species having the same general habits, occupying analogous situations, and obviously filling the same place in the natural economy of this archipelago, that strikes me with wonder. It may be suspected that some of these representative species, at least in the case of the tortoise and of some of the birds, may hereafter prove to be only well-marked races; but this would be of equally great interest to the philosophical naturalist.

That same year, Wickham, sailing along the north coast of Australia in the *Beagle*, spotted a likely port and named it 'Darwin' after his friend.

When Wickham's sailing days were done, he settled in Moreton Bay as the penal colony transitioned from housing convicts to free settlers. Wickham became Brisbane-town's 'First Resident' and its first police magistrate. He moved into a stately home on the confluence of the river and Breakfast Creek, Newstead House. But Darwin's tortoises were unhappy in England. The climate didn't suit them. What better place for his tortoises to live, Darwin thought, than with his friend John Wickham in subtropical Brisbane?

And so when *On the Origin of Species* was published in 1859, inspired by Darwin's Galapagos tortoises, these great chelonia were living quietly in the gardens of Newstead House at Breakfast Creek in the distant Antipodes. As the theory of evolution upended what we knew of ourselves and our planet, Darwin's tortoises plodded through their days in muggy Brisbane. When Wickham ultimately left Brisbane-town for France and, soon after, a grave in Biarritz, the tortoises were given to the zoo at the Botanic Gardens. By the time the zoo closed in 1952, only one survived, Harriet. She was sold to a famous naturalist, David Fleay, who in turn passed her on to another naturalist – one with a sharper eye for a public stage – Steve Irwin. Harriet died in 2006, aged 175, a living link to Darwin and his revolutionary, evolutionary theory.

Only maybe the ancient tortoise wasn't Darwin's at all. Maybe the documents that could prove Harriet's provenance don't exist. And maybe it was just that the gaps in her record were big enough to fill with some Darwin stardust.

Old she was. Mitochondrial DNA testing in 1998 revealed she had been old enough that 175 years at death is as good an estimate as any. But that same testing also cruelled hopes of proving her provenance. Harriet was a *Chelonoidis niger porteri*, a subspecies that lived only on the island of Santa Cruz, and Darwin's tortoises had been gathered from other islands – Española, Santa Maria and San Salvador. Definitively not Santa Cruz.

The written record? None of Darwin's letters or notes or articles refer to him having given his tortoises to Wickham. And while Harriet's own history might be traced back to the Botanic Gardens, there are no longer any public records from her younger days to tell us when she arrived and how. Because – and I love this – the river intervened. The flood of 1893 which devastated so much of the young city took the records of the Botanic Gardens' early years with it, sweeping away all detail. Rivers record history in grand sweeps. They work on a mythical scale.

~

We pass an ancient river gum that must have toppled in the flood. Its dark, Medusa-haired root complex is now exposed to the sky. What to make of all that secret knowledge now revealed?

We find some shade for lunch. True to his nickname, Piglet sinks into a bog, exhausted, seeking the cool of the mud. He stirs when Steve peels the plastic sleeve off a salami roll, though won't take the slice Steve cuts for him, at least not from Steve's hand. Steve rests it on a log, and only then does Piglet take it and move away. We untie our bootlaces, lean back on our packs, close our eyes and drift.

The river widens. Linville is still some kilometres away on the right bank, set back from the river by a couple of hundred metres. We decide to stay on the right-hand side of the river for the remainder of the afternoon, in case the water is too deep and too wide to wade back across as we near the village.

By three o'clock we're picking our way through lantana on the high bank. After thirteen kilometres Piglet seems indefatigable on his little dachshund legs. Below us the water is clear enough to see the gravel on the bed. The land contours back down to the water, where we find a pump with a ten-inch-diameter pipe – an enormous water-pumping capacity. The pipe runs up and over the hill, where to we can't see, but there's the faint chug of a distant engine, and we guess there's a farm above us somewhere.

Just beyond the pump, a large rocky outcrop leans over the river, blocking our way. The only route forward is through a cleft between the outcrop and a towering cliff. A small-leaved fig stands sentinel atop the great rock, its tentacle roots wrapped around the stone as if fused to it, steady, immovable. This is the way, the fig seems to be signalling – come this way. We slide through the crevice and up to a saddle between the outcrop and the high rock face. From the saddle there is no path ahead down to the river. But we spot what we think is a route up the cliff-face, one that follows a fissure in the rock.

Steve leads the scramble. I follow at a safe distance. The route up the fissure is a puzzle to be solved. A pair of sturdy saplings that have taken root in the cliff provide leverage. I look for cracks in the rock offering potential as a foothold – wide enough for a boot to get purchase, without becoming jammed stuck. I find handholds on the edges of rock or hook an elbow around the trunk of a gnarled fig. But I'm tiring. I want to catch my breath, but I press on. My legs are heavy. My forearms burn. I'm dragging myself up hundreds of millions of years of rock. The weight of my pack pulls at me. It's as if my tent and sleeping bag and the food I've not yet eaten and my journal and first-aid kit and spare batteries and water – the very provisions necessary to support this trek – are not just a burden, but are themselves the danger. I pause and pant and gather myself. At what point should I unclip my pack and let it tumble away into the river below? But the tree line at the top of the cliff is only another twenty metres above me, and if I just focus on the next metre, and then the metre after that, soon enough I'll get there.

I join Steve at the clifftop, where he's catching his breath beside a fence post. For a moment I feel like that fig on its outcrop, as if I've been somehow transformed by the hundreds of millions of years I've glimpsed. That my pounding heart has found the pulse of the earth itself. But then I see behind Steve, on the other side of the fence, the strangest of sights – at least in that moment – the well-tended rows of an avocado orchard. The engine we'd heard at the pump is the throbbing of a tractor at the far end of the orchard. It's not just the alienness of the avocados out here that's disconcerting. It's the neatness of the plantation after the primeval cliff.

But Steve's had time to collect himself. And to realise the thing we'd both forgotten in our focus on the cliff.

'Piglet,' he says.

I look down, and there he is at the base of the cliff, scrapping this way and that, anxiously seeking a way up. But dachshunds aren't built for scaling cliffs. Our eyes meet, and I feel a terrible tug at my heart.

We consider the options. One of us could scramble back down the cliff with an empty backpack. Perhaps we could persuade Piglet into our arms. Perhaps even into the pack. But the descent of a cliff like this is always more treacherous than the ascent, and whatever Piglet's history and nature, he's not going to allow himself to be tucked into the pouch of a backpack. Even if we could, the risk of him squirming suddenly in the pouch on the return climb and causing whoever had him on their back to topple … it's not worth it. Another option is to try to find an alternative – but inevitably longer – dachshund-friendly route. Steve volunteers to look and heads off along the clifftop, making his way back upriver, disappearing among the trees. I gulp down water while I wait. Steve soon returns. There's a large shed just out of view, and the sound of men working on machinery inside, and there's no way around without announcing ourselves.

I'm fairly confident I've spoken to all the landowners on this stretch of the river. Fairly, but not completely, and the truth is I don't know whose land this is. If I'd seen this orchard on a satellite map in my planning, it hadn't made an impression. We look at each other. Half-an-hour has passed since we left Piglet. We can no longer see him at the foot of the cliff. Even if we found another route up here, the dog might already have headed for home. We tell ourselves that Piglet is a resourceful dog and won't have any trouble following the river home. It's part of the genius of our species that we can reassure ourselves of just about anything.

~

The reward for our cliff-scaling is a labour even more demanding.

We skirt around the prim orchard, the sound of the tractor fading as we put distance between us and it. The property on the other side of the boundary fence is, unlike the orchard, heavily grassed. It's obviously not carrying cattle. We plot a course through the grass towards a rise from which we should be able to take our bearings again. Steve and I have both been around enough snakes over the years to respect the browns that like this sort of country, and the memory of the red-belly black that crossed our path a day or so ago is still fresh. We'll be pushing through chest-high grass unable to see what's at our feet. What we have in our favour is that our body-bashing will be loud and will give snakes plenty of time to flee. Even though we're both wearing gaiters, I'm in trousers, while Steve's only in shorts. It's my responsibility to lead.

We take on the grass, leaning into it and using our momentum to drive our bodies forward. We have to trust that where we're stepping is even. But as the ground starts sloping towards a concealed gully, we slow, careful not to misstep and fall. As we lose pace and momentum, each step takes more effort. Thankfully the gully is just a gentle fold in the terrain; soon enough we're across it and start up the opposite slope. There's no thinning of the grass. I press into the ascent, Steve following in the path I leave behind me.

The grass gives way to lantana. I stop to catch my breath and wipe sweat from my eyes with my shirtsleeves as I map a route through the lantana to what I hope will be the top of the hill. We thread our way between the lantana bushes as if we've found a corridor through a labyrinth. Gradually the path I've chosen narrows and then closes in around us until we're facing a lantana wall. We could backtrack and look for an alternative route, but the top of the hill is close. Deceptively close.

I plough forward, stamping on low branches and using the mass of my body and pack to crash through. Stepping over other branches, wrangling a route over and through, sweat pouring off me. The lantana catches my hat again and again. Again and again I stop to retrieve it from the end of a branch. The scratches on my cheeks and neck sting with sweat. Up ahead a lone gum stands tall, reaching high above the lantana jungle. I feel myself veering towards it, some solidarity between us in the face of a common adversary. Then I check myself. The lantana will have woven itself tightly around the gum, will be even thicker around the base of the tree. I reset our course.

The going gets harder. I'm barely thinking now, just struggling forward, wrestling a way through step by torturous step. I start to falter. I've been losing water faster than I can replenish it and feel myself starting to dehydrate. Lunch was a long time ago now, I'm burning energy fast, and this lantana is as thick as I've ever encountered. Though it's not just lantana on this last stretch to the top of the hill – the lantana is laced with thick vine. Eventually – inevitably – I fail to lift my boot high enough and the vine catches my ankle, and I pitch forward. The lantana is so dense that I lie suspended in it. I stretch out with my right hand but can't touch the ground to right myself. The more I struggle, the more the lantana branches find new ways to grip me, wrap round my pack and tighten their hold. I roll onto my side and relax and wait for Steve to reach me and pull me up.

From the hill we see a fence. On the other side the lantana thins, and beyond the scrub we think we see an overgrown vehicle track descending towards the river.

~

The riverbed on the last kilometre or so into Linville is the widest we've encountered, a hundred metres and more bank to bank, a mainly dry bed of stone and sand, the narrow channel of water hugging the far bank. Walking down the middle of the sandy bed, we're following the current as surely as if we were in a canoe. The bed ripples forward in what seems like an endless series of gently sloping sand ridges. Swells of grey-gold sand rising upwards to a lip, then falling away to a deposit of blue pebbles at its foot. We're walking in flow, in the heart of the moving river, propelled along by waves of pebble and quartz and sand. The regularity of it is mesmerising, like a musical beat. We fall into the rhythm of it, the ancient flow of water.

~

Just after four o'clock, we step into the public bar of the Linville Hotel, sweaty, scratched, dirty, sore. Spent. The hotel is an old two-storey, wide-verandahed timber pub on the main street of town with a stylised stag's head and antlers as its motif. I slough off my pack. I know Steve will have a beer, because we started talking about it a kilometre back. Tracey behind the bar waits for me to step forward. She guesses we've booked a room. 'We need a beer first,' I say. It's an unoriginal line, but the locals at either end of the bar are forgiving.

'Where have you come from?' asks a bloke in a singlet sitting on a stool.

'Up the top of the river.'

'And where are you going to?'

'Brisbane.'

'You'll need more than just one, then,' the woman sitting beside him says dryly as Tracey places schooners in front of Steve and me.

We drink that first beer quickly, proving the woman right. But a second beer also gives time to try to explain, in response to their genuine interest, what this walk is about.

A trek from the start of the Western Branch along its banks to Brisbane and then Moreton Bay. To keep the river in sight as much as it's possible to. To keep it companion. That Steve and I have walked the first five days, and tomorrow he'll leave and my son Dominic will join me for a three-day stretch. How long will it take? I don't know for sure. I'm guessing four weeks, but I have another week set aside if I need it.

'What about the weather?'

Steve and I have been completely disconnected from daily news since we left.

'What about it?' I ask.

'The forecast is for rain.'

I shrug.

Tracey shows us to our rooms upstairs, where our gear is already waiting for us after Ali dropped it off this morning. Tracey and her sister Tanya and their husbands are the new publicans. The sisters are fifth-generation on the river, and have returned home to run the pub.

~

As I unpack, my phone rings. A man introduces himself as Ali's husband.

'You wouldn't happen to have seen a spotted dachshund today, would you?'

That terrible catch in the gut when suppressed fear becomes reality.

I explain what I know – how four dogs followed us, that we'd tried to send them back home at the second causeway, but that the dashie wouldn't go with the others. That we'd decided that if he accompanied us all the way to Linville we'd leash him when we got to the pub for Ali to pick him up tonight. A pointless self-justification. I tell him his dog stayed with us until the cliff.

'Which cliff?'

I try to describe where it was, but my efforts are hopeless. I'm a stranger in this country.

I almost say, 'And that's the last we saw of him,' but instead manage, 'We figured he'd head home after that,' though there'd been more hoping than figuring.

'That was a couple of hours ago,' I add.

'Right, then.' Ali's husband pauses. 'Well, I think we'll just have to wait for him. We've only had him for a couple of weeks, and he hasn't yet quite attached himself to us.' For my benefit, he adds, 'I'm sure he'll find his way home.' Ali's husband is a decent man.

Steve and I feel sick. I distract myself by writing in my journal over a beer on the deck downstairs, Steve by playing Galaga on a tabletop arcade machine. I drift over to watch him defend himself against armada after armada of alien spaceships. Galaga was my favourite computer game for a while when I was a kid, a game that had, back then, arrived from a boundlessly unknowable future and is now a relic of yesteryear. A time of clarity about what was alien and what was not, what to kill and what not, who to love and why.

My phone rings again.

Steve looks up at me from his game, aliens pouring in through his temporarily abandoned defences.

It's Ali's husband.

'He's home.'

Oh, the relief. I give Steve a thumbs up and smile. A neighbour had found Piglet on the river road – how had he got there? – recognised him, stopped and gave him a lift back in his truck. Ali's husband will see us soon for dinner.

~

It's not just Ali and her husband who join us, but Graham. This is the first time they've met. Graham knows the history of every property on the upper Brisbane river since white people arrived in the district, though they're not chronologies as much as testaments of ambition and loss. He knows more about the history of these farms than most of their owners. It's his duty to keep this history. Ali and her husband hear about who owned their place before them, and who before that. Graham compliments them on their fences, and means it.

I recount the day's events. Having bumped into Bruce, who of course they know. Sharing the day with Piglet. The deer. The cliff and the avocado farm. We learn who owns the farm. I then casually observe how overgrown the next property was.

A silence.

'That's my place,' Graham says. I am deeply embarrassed. On two counts: at the implied criticism that he'd let his property grow out of control, and also because I'd visited it the previous year when he was showing me around but had no sense today, as we crossed it, that it was the same property.

Graham has recently bought back onto the river after some years in the semi-arid west of the state. He's clearing this block to put cattle on it. Because this is cattle country, and he knows cattle. And because he's returning to the river, coming home. Last year when he and I entered the property it was from the river road. We'd opened gates, and driven his LandCruiser along a newly cut track to a view from the top of a ridge. But today, coming from the river, on foot, his property had been unrecognisable. It had nearly defeated us.

~

There's a language – almost their own – that these river people use, a language woven through generations of living in the upper reaches of this river valley. Of cattle and tree felling and rutting deer in the hills and being cut off from the world by flood. When the river runs after rain, it's a 'fresh'. A deeper section of the river is a 'hole'. The antlers a deer drops annually are 'cast-offs', and the deer, mainly, are 'reds'.

~

My own family – the Cleary line – grows out of the forests and streams and bays of Ireland. A land where goddesses dwelt in rivers. The Shannon was formed when Sionnan went to a well to catch the salmon of wisdom. But seeking eternal knowledge is often forbidden, and the walls of the well broke. Sionnan was drowned in the deluge that rushed out to the sea, creating the river in which she still resides. The O'Cleirighs were bards and scribes, history keepers from Connacht and later Tír Chonaill in what is now Donegal. Farmers too. When they left Ireland and made their months-long ocean crossing to this continent, they disembarked as migrants and left the coast immediately, heading inland. They moved onto Jagera Country, Yuggera Country, Ugarapul Country and Wakka Wakka Country, and then my grandfather moved to Crows Nest and later to Toowoomba on the range, Jarowair and Giabal Country. And in time I have rolled down the range to Brisbane on the river.

So I am not of the river like James and his family, or like Graham, Ali, Bruce and Marjorie are of the river. For them the river offers boons and cruelties and wisdoms daily. It flows through their blood and scores itself on their skin.

And yet I am also of the river. I too reckon with it every day. Without it I would not recognise myself. I live in the city through which it flows. I cross it or spy it or walk beside it or run along it and breathe its river-filtered air. I live and work in buildings erected from sand and gravel extracted from its bed or its alluvial plain. I drink it and wash in it.

Now I lie in my anonymous hotel bed, where yesterday a travelling salesman lay and the day before that a ganger and before that a touring cyclist or a roadworker or the cousin of a bride come to town for a wedding. There's a breeze blowing off the river. It has blown for a million years and more.

Day 6

Linville to Moore

Distance: 9.52 km

Evening Camp: Place of the Moreton Bay chestnut grove

I don't know if it's the crowing rooster or the rain on the tin roof that wakes me. It's dark outside, and the rain is insistent. I turn on my phone and check the forecast. It's Friday morning. The Bureau of Meteorology's website says there's a ninety per cent chance of rain every day for the next seven days, with forty millimetres predicted for next Wednesday and Thursday. I drift back to sleep, resigning myself to the likelihood that a week of cloudy night skies may mean not having the chance to point out the Dark Emu to my son Dominic when he joins me for the next leg of the journey.

~

Steve's wife, Nic, arrives for breakfast on the verandah. We have bowls of cornflakes and mugs of instant coffee, and Nic knows precisely what to ask about the five days of walking: enough, not too much. They will leave together and drive back up the river road, and Steve will show her where this happened and where that did, and where the skull and antlers of a great stag may still be resting where he'd left them.

We embrace on the verandah, and I miss him already. I walk around to the street side of the hotel. Steve and Nic appear again below, and we wave and they climb into their car and soon enough they are gone.

I see, then, another vehicle parked in front of the hotel, a huge logging truck – a B-double with its two flatbed trailers. The trailers are empty. Except, as I examine them more closely, they're not. The wet trays of each trailer are strewn with bark and timber chips – traces of trees cut from a forest on a nearby hillside. Surely plantation timber, I tell myself, surely not old-growth. I think of the plaque from a day or so ago commemorating the men who'd built the roads from this village to the head of the river so loggers could get to the hoop pines that grew on the highest of the valley's hills. But now? Now the forests are state forests. But from which forest exactly did this truck gather its logs? Which hillside? Which mill have they been delivered to? And what, ultimately, will become of those trees?

~

The old-growth hoop pine forests are long gone, or nearly so. *Araucaria cunninghamii*, after Allan Cunningham, the first European botanist to explore the river and collect specimens. He didn't get this far up, but he was part of the third European expedition up the river, with explorer and surveyor and Brisbane city founder, John Oxley, in September 1824. Hoop pine was always a likely timber. Tall, straight, soft, knot-free. In the wild, a hoop pine can live for 450 years. In a plantation, they are turned around every forty-odd years.

'Is there any old-growth hoop pine left?' I'd asked a riverman I spoke with before I set out, the son of a logging family who'd once cut hoop. Back then it was a major operation: teams of men working their way over the decades through increasingly rugged country, using fantastically long pine chutes to slide logs down from the top of a hill, sometimes – particularly further downstream – directly into the river, where they'd be floated to the mill. Other times the timber would be carted to railheads like Linville before being transported away.

'There's still some old hoop up there,' the riverman answered, 'up the very top of the valley.'

It sounded like he was talking about some mythical creature. He still cuts timber himself: ironbark, spotted gum, box. If he cuts, he does it privately and uses handsaws, not an excavator. 'It's a trade-off,' he said. 'Using an excavator is safer, but a handsaw allows you to pick with more precision and doesn't do as much damage to the canopy when the tree falls.'

The floorboards of my home, built in 1929, are hoop pine. They could have been cut from up here. Perhaps I walk barefoot, every morning, on some of the last old-growth hoop to have lived on these upper Brisbane River hillsides. I might, unknowingly, be the beneficiary of their felling. The weave of past and present, the untraceable threads of community, so many invisible connections.

~

My wife, Alisa, and my son Dominic should be arriving any minute. Dominic is eighteen, fit and strong, and a great walking companion; last summer, after he finished school, the two of us hiked for days in the Australian Alps. I look out over the handrail at the day. The rain is holding off, or pausing. It's grey, but it feels like the cloud might be lifting in the east, riverwards. I notice a colony of flying foxes in a stand of blue gums by the river. Seeing them, only now do I hear them as well. I'm too sight-dependent. So I close my eyes and in time the colony separates into individuals. This one screeching, and then that, and then another.

It's midmorning when Alisa and Dominic get here. Alisa and I embrace halfway up the pub's wide stairs and laugh, the site of our reunion so clichéd it could look staged. A week ago – on the eve of setting out – she'd predicted my fifty-three-year-old body would let me down. She wasn't entirely joking, though she was really just offering cover: if you have to stop, it'll be because of your body, not you. Alisa's glad my body's holding up. Then Dominic and I hug each other, father and son and the excitement of the unknown ahead.

This is the first changeover of walking companions and therefore also the first resupply of provisions. On the verandah outside my room I treat my blisters and fill my backpack with fresh supplies: dehydrated meals, an apple, an orange, an avocado, a carrot and a cucumber. I stow extra blister packs and swap five days of dirty socks for clean. For the next leg of the walk, Dominic and I will share the two-person tent he's brought with him, so I temporarily shed the one I've been using. While I'm busy arranging my gear, Alisa discovers that one of the publicans downstairs lives half her life near our home in Brisbane. She's both a daughter of the valley and a city accountant – perhaps it's possible to love a river and a city equally.

~

It's eleven o'clock before Dominic and I say our goodbyes to Alisa and leave, late, but that's okay, because it's a short day, ten kilometres. We take the right bank, the default bank for the journey. The bed is sandy and wide, and easy walking.

After a kilometre or so, we see a couple in fold-out camping chairs fishing beside their four-wheel drive, cans of beer nestled in the arms of their chairs. Their two fox terriers scamper towards us, bundles of yap. I call out a greeting over the noise of the barking dogs.

'Do you have the owner's permission to be here?' the woman responds.

'Yes,' I say, confidently, though I'm not sure precisely whose land we're on.

'Oh, really,' she replies, 'you know the owner?'

I tell her about the walk, and that I've spoken with over a hundred owners, who've been happy for us to walk through. She looks at me, then Dominic, taking us both in. She turns back to the river and says, as much to the fish she hopes to catch as to anyone else, that her friend gets a bit touchy about people walking on his property, but if we stay down near the water she doesn't think it'll be a problem.

Once he's satisfied his wife's interrogation is over, the man tells us they're after catfish – not fork-tailed catfish, usurpers in these waterholes – but 'real' catfish, by which I think he means the eel-tailed variety. They've caught a couple, he says, and by now the woman too is happy to chat. She tilts her bucket towards us to show us their catch.

As we continue on our way, the river narrows. Something is happening up ahead. Vegetation – bottlebrush and then a thickening patch of hip-high castor-oil plants – crowds out the bed, until suddenly we find ourselves confronted by a creek pouring in from the right. A large creek. A creek stirred up by the rain, thickly brown with soil and silt washed into it overnight. It's impossible to tell how deep it is, but it's dark and wide with steep, muddy banks. There's no crossing this one, indeed nothing to be done but to track the creek upstream until we find somewhere to get across. It's the first time I've had to leave the river to continue the journey, and it hurts to turn my back on it and start walking away, even temporarily.

The steep bank is not just muddy – it's slippery as glass, and deeply shaded. This is – I can't quite believe it, am unprepared for it – rainforest. Dominic's boots are older than mine, with wearing tread, so I pass him one of my walking sticks. We drive our poles into the mud and make for a Moreton Bay chestnut tree. We slip and slide and hug the first tree we reach, catch our breath and then strike out towards a second, hauling ourselves diagonally up the slope by grabbing hanging lianas, step by unsteady step to the top of the bank.

When we emerge from the strip of forest into a paddock, we can see the guardrails of the valley road up ahead and, a little way down the road, a simple bridge over the creek. *Greenhide Creek*, a signpost on the approach tells us. We pause on the bridge, bitumen underfoot, and look down at that formidable creek. We step back off the road as soon as we're across. The remnant rainforest is wider here, deeper. A sign says *Greenhide Scrub*, but to call it a scrub feels inadequate, almost a slight. In the days when forests like these were obstacles to farming, the naming was probably deliberate. This forest may only be a few acres, but it is amazing. Ecosystems are collapsing all around us, but not here, not yet, not now. This resistance is fierce.

We find a track heading through the forest down towards the river. We wend our way through more Moreton Bay chestnuts, ashes and a patch of cunjevoi (*Alocasia brisbanensis*) before I spot a red cedar. What is it doing here? I think. By which I mean, how did it survive the logging? Dominic lifts his head to gaze at the canopy above him. If that's not prayer, what is?

Then we spot signs of more recent human activity: of plantings – dozens of young natives protected by shin-high, pale green, triangular corflutes – and an understorey cleared of invasives. These patches of rainforest are described as 'remnants', but they're also gardens, lovingly tended by volunteers: believers, warriors, visionaries.

What did this country look like? That question from the first day, before we set out. That question nagging so many of us. In parts it once looked like this.

~

The sun has burnt off the cloud by lunchtime. We find a riffling stretch of fast-flowing shallows and strip and lie on the stony bed, immersing our bodies in the coursing water. Water that's come out of the Western Branch and the Eastern Branch and Cooyar Creek and from swollen Greenhide Creek just an hour's walk upriver and another half-dozen upstream tributaries. It's hard to reckon. An hour ago Dominic and I were desperately trying to arrest our slide down a muddy creek bank, and now we're carefreely skimming pebbles downstream towards Brisbane, weeks away.

~

After lunch we stride out, refreshed, taking the left bank. The bed widens, a fifty-metre-broad beach of pristine sand and gravel. Two sea eagles leave their purchase high in the branches of a dead gum and begin circling, flaunting their magnificence against the canvas of a willing blue sky. Wingspans large enough to obscure the sun, pure white breasts, darkening wings. My son and I walk the river while the birds circle and sweep above us, and it feels in this river-blessed moment that we are entirely welcome.

We talk as we walk, roaming from topic to topic, one flowing into the next as if there are no boundaries between them: how different this river is from the alpine Snowy River we'd traced for days in the summer just passed, the genius of Kendrick Lamar's music, which university subjects Dominic is enjoying and which he's not, the latest crisis in American politics, how high a sea eagle can fly. My son asks me about his grandfather – my father – who died when he was seven, keen to add my stories to his memories.

'Dad loved taking us on bushwalks,' I say. 'If it wasn't for him, we wouldn't be here now. Though only ever day walks. We never camped overnight.'

'Why not?'

It's a good question. I should know the answer, but I'm not entirely sure. I start thinking aloud.

'Maybe the logistics were too much with five children. But

also … I don't know … he was … a busy person. He wasn't the sort
of man to spend hours looking into a campfire.'

'You should talk more,' Dominic says. 'You've got takes on things,
but too often you keep your opinions to yourself.' Ah, these walking
truths.

We talk our way forward and then talk our way back in time,
shifting from experience to aspiration, from knowledge to doubt,
traversing territory old and new. The river is the third participant
in the conversation, laughing over shallows, burbling round bends,
falling silent as it widens and slows, offering gentle lessons in the art
of conversation.

A northerly starts blowing. We crunch our way over pebbles. We
step up onto a cattle pad and follow it for a while. A Charolais cow
stands silently and watches us pass. We fall back into the bed where
it is sandy and, after the pressed earth of the cattle track, enjoy the
way the sand gives beneath our boots.

The flowing water draws us near once again. We veer towards
the corrugated river and spot the first live turtle of the journey,
bouncing up and down on a log that's fallen across the stream. The
turtle's bouncing looks deliberate, as if it's trying to generate enough
momentum to dive, at the top of its lift, into the water. I want
to identify it – is it a saw-shell, like the carapace from a few days
ago? – but before we get close enough, it slips off the side of the log
and plops into the river. I keep my eyes trained on the bank as we
walk on, examining each log and boulder, hoping to spy another.
We walk deeper and deeper into the afternoon, but I can't find one.

In front of us a low bridge appears, Allery's Crossing, the one
we're aiming for. To the right, over the river, there's a house on a hill,
and below the house, on the bank, are a huddle of abandoned timber
sheds, a stand of Moreton Bay chestnut trees and a clearing – our
campsite for the evening.

~

Kelvin is working his cattle in the yards on the other side of the bridge that bears his family name. The house on the heights is his, newly cut into the hill. He sees us from his tractor when we're about halfway across and raises his arm in greeting. We raise ours back.

'Need a hand?' I ask when we reach the yards. 'The young fella's looking for something to do.' I wink at Dominic.

'Nah. All good. All good,' Kelvin says. 'Got some weaners in yesterday. Had to get 'em in yesterday … The sale's not till next weekend, but …' He looks at the sky, which is clouding over. 'With the forecast, the rain'd make it impossible to get 'em in next week. I can't afford to miss the sale … so had to get 'em in. Problem is, there's no shade, so I need to sling up some tarpaulin before the rain starts.'

Above the heads of the twenty weaners milling in the churned mud are large sheets of taut grey tarpaulin, neatly strung up between a high railing, the roof of an adjoining besser-block shed, and a fig tree.

'Sure you don't need a hand with anything?' Dominic asks.

'Nah, go and set yourself up, but I tell you what – why don't you sleep in one of the old huts?'

He points at the abandoned buildings across the road.

'If it's pissing down, you're better off under there than in a fucking tent, aren't you?'

We all laugh, but he's right. We don't particularly want to be stuck in a tent in a storm.

'Seriously, push the hay aside, and set yourselves up so you don't roll over the edge in your sleep. Anyway, go and have a look.'

We have a look. The huts were once dormitories used by schools for outdoor education camps until they were abandoned in the 1980s or 90s. They're raised off the ground on concrete stumps, but are shells: the sheeting of the internal walls has been removed, with only the frames left, and even some of the external walls have been pushed out entirely. Only shards of glass remain in the casement windows. Sparrows and willy wagtails have moved in. Cattle shelter beneath the huts and prickly pear grows on the roof of the vehicle shed. However the floor of the largest hut is solid, and the corrugated iron roof looks new. It would be easy to lay a sleeping mat and bag down. And there's a superb view down to the river through the gap where a wall once stood.

But the flat by the river is as good a camping spot as you could ask for. It's well grassed, there's an easy track down to the water, and the stand of ten or twelve tall and handsome Moreton Bay chestnuts is very, very beautiful. Kelvin and his wife got married down here one spring when the chestnuts were flowering their big, bold red and yellow blossoms.

We decide to try our luck beside the river, but we'll decamp to the hut if the weather gets ugly. Someone has been down here recently and left half-a-dozen chainsaw-cut stumps as seats around a charred circle. After pitching the tent, we gather leaves and grass, twigs and sticks. I snap larger branches in half under my boot and drag them down the hill. Everything is wet. As I'm arranging the firewood in the last light, there's movement on the water behind Dominic's shoulder and I look up. Oh, my. Two black swans are moving silently down the river. We stop and sit and watch the swans glide their way into the night.

~

'Just keep feeding it,' Vic says, leaning forward to push the unburnt part of a piece of wood deeper into the sputtering fire. Vic is an old riverman I'd been trying to contact for weeks, who's lived his life in the upper valley. His wife had passed on my message that we'd be camping here tonight, and he'd seen our fire from the road on his way home after the last of his day's errands. 'It's damp, but if you just keep ...' He trails off as he reaches for one of the clumps of grass

Dominic is drying beside the flames. 'Yes, just ...' It's his hands that finish the sentence, coaxing a fire out of the wet wood.

An inch-and-a-half has fallen in the last couple of days, he tells us – like most farmers, he keeps a rain gauge – 'But we've got this going nicely,' he says. There is nothing more important than the fire, and Vic took responsibility for it as soon as he pulled up a log and joined us as darkness fell. He may be nearly eighty, but he is slim, and fit, and spry, and generous with his story.

His grandfather arrived in the valley after the First World War and his family stayed on. Vic's got a couple of properties on the river, here – he points across to the other bank – and closer to the headwaters. Steve and I had walked through one of his farms a few days ago. He tells Dominic and me about the river. About his family. About a time before the dam – Wivenhoe, still a week or so walk downriver – which changed everything when it was built in 1984. He lays out a history of this stretch of river by reference to the floods. He lists them off. The one in 1955 was a major flood – thirteen inches in eighteen hours from an inland cyclone. There was 1974, of course, the worst in his lifetime. Then 1999, when a third of the bridge was washed away. In 2011, two surges, twenty-four hours apart, caught everyone off guard. Then 2013, which no-one in Brisbane talks about, but which was only a metre below 2011. And now February 2022.

When it's in flood, he tells us, the river is three-quarters of a kilometre wide here. Imagine that.

'These major floods ...' He reflects for a moment, feeding another branch into the fire. 'It used to be narrower, the river. It always had a lot of depth. But '74 ripped the guts out of it, and then '11 and '13. Having those two floods straight after each other, it ripped the banks about. The river takes gravel from one hole and deposits it somewhere else. A flood changes the course of the river. Kangaroo Creek – you'll see it tomorrow – Kangaroo Creek used to come in at a right angle, but 2011 took a whole lot of gravel down the creek and dumped it in the river. Now the creek comes in fifty yards further down.

'You can't pin a river down, not in flood, not anytime, not really,' Vic says. 'Over in America, when they were doing cultivation on the Mississippi, they tried to cut it square, see. They tried to make the river straight, but a river doesn't like doing that. It gets too impetuous. So what it does is, it starts chopping off a bank because it wants to slow itself down. And I think this river does the same.'

The smoke starts to follow us. We stand to get our faces clear of it, to reposition our logs, then feed the fire some more.

'But don't forget drought. Drought shakes the river up too. See that hole?' He's pointing out at the swiftly flowing river over my shoulder.

'Hole?'

He sees something I don't.

'Not long ago, in 2009, maybe, or 2010, that hole had very little in it, was almost out of water. The river wouldn't flow. There was just a series of holes from here back to the headwaters. People were getting dozers in to clean out the gravel from their holes to try and get down to the water level. They'd do it once or twice, but then they'd get to bedrock.'

The basalt Steve and I had followed.

'People around here who'd never been out of water were digging and scratching. Sometimes it's just a matter of hanging on.'

Day 7

Moore to Harlin

Distance: 15.23 km

Evening Camp: Place of the three river channels

It didn't rain overnight, but the cloud-cover this morning is complete, grey and threatening. We break camp and, as Vic suggested last evening, cross back over Allery's Crossing to the left bank. There's a cattle pad, recently overgrown with lantana. We shoulder our way through, wrestling our packs free when branches catch and try to hold them. The track rises and the lantana thins and we find ourselves high above the water. A red stag disappears over the top of the hill. Ahead, a sea eagle takes wing. Is it one of the same birds as yesterday?

Opposite, the recent floods have scoured the bank bare, revealing the layering of thousands of years of sedimentary deposits.

Kangaroo Creek is just as Vic described it. We crunch over the mounds of gravel where not that long ago a creek mouth fed the river. At the newly positioned downriver mouth, it's shallow enough to cross without taking our boots off. On the other side, behind a screen of callistemon, we hear a burst of water-slap. Three Pacific black ducks launch themselves into the air and fly further downriver.

I've been walking along the river for seven days now and am beginning to see it differently. I look ahead with different eyes, scanning the bed and the banks, taking in the cut of the water and changes of vegetation, transitions from pebble to gravel to sand, half an eye on the best route. Because some routes are easier than others, and choosing a difficult path has consequences. So we stay on the inside bend where possible. Another pitfall: what might appear to be a low bank may become, as we follow it obliviously downriver, a finger of dry land bounded by the flowing river on one side and a concealed lagoon on the other. Sure, backtracking is a part of life, but when you're tired and hungry and looking forward to dumping your heavy pack as soon as possible at the evening's campsite, a long backtrack is disheartening.

Dominic and I cross the river, boots and socks and gaiters off this time. On the other side we sit on logs, munch our muesli bars, and wait for our feet to dry.

The very next creek, when we come to it, is named Emu Creek. This used to be kangaroo and emu country. Even the white settlers knew that. It's become a refrain, nagging away, returning, revealing. Kangaroo and emu country. Emu Creek is not to be doubted. It's wide and is emptying a couple of days of run-off into the river, and once again we'll need to make our way up the creek to find somewhere to cross.

We head inland, and as we round a clump of lantana we find a Brahman bull on the bank a few metres ahead of us. We stop, my heart racing. He startles too. Big as he is, Brahmans are usually pretty docile, and Dominic and I – standing close together, backpacks bulking us out even further – have enough presence about us to make him turn and head for the river. Judging from the height of the water over the hump on his back as he crosses, it'd be chest-high for us if we tried, and the force of the water would sweep us away.

There's a concrete ford a couple of hundred metres further up-creek where a wooden railway bridge once carried trains to Linville and then even further up the range to Yarraman. Remnants of the bridge – wooden piles and lengths of iron, sleepers with rusting bolts and square washers – litter the creek, pulled out and down and away by floodwaters. We wade across the ford, and once we reach the other side, dry our feet and lace our boots and make our way down the creek to the river.

The weather starts following us, coming in from the northwest.

A pair of black swans, mates for life, travel downriver at about the same pace as our walking. The river water, with Emu Creek's run-off pouring into it, is changing colour. The dulling sky accentuates it – muddy brown. We pass a large river gum. A rope swing hangs from its branches. I smile as a lifetime of memories of kids jumping from ropes into rivers sweeps over me: schoolfriends, Dominic, my other son, Liam, their friends, nephews and nieces. All the thousands of kids I'd watched somersaulting off banks or cliffs as I paused beside rivers the world over. All the rivers I've swung into myself, timing the release just so, all those airborne moments. All those river embraces.

We sit under the branches of the swing tree for lunch.

Eventually the weather catches us, and it begins to rain. We fix covers over our packs and pull the drawstrings tight. We don our waterproof jackets. Ahead the river is wide, the bank steepening, shifting into hill. We're stuck on an outside bend and will need to leave the water to navigate round a sweeping cliff. We follow a cattle pad as it angles up the slope, rainwater beginning to course down the pad, mud becoming muddier. The grass at the top of the hill is high. As we catch our breath, we gaze east across a paddock of even higher grass to a gravel road that will take us back to the river.

We wade through chest-high red Natal grass towards the road, rain falling lightly. The sky is wet, the grass is wet, our clothes are wet and heavy. A group of quail bursts from the grass at our feet, startling us. We laugh at ourselves above the sound of the rain. The bottom strand of the barbed wire fence running against the road is high enough to roll under, commando style, without needing to take off our packs. Down the wire a green lorikeet remains perched on its fence post.

It's good to be out of the paddock and walking on a gravel road, unlatching a gate rather than climbing over it.

Last year I drove through this gate and down to the end of the road and onto the bed of the river to learn about fingerlings, and an attempt to repopulate the river with a cousin of the extinct Brisbane River cod. Now – after February – it'd be impossible to get a vehicle onto the riverbed because of the sheer metre-and-a-half drop from the end of the road. The river's grand erosive work really can be breathtaking.

Not that the river cares much about what I might think of it, about what any of us might think. After all, since at least since 1543, when Nicolaus Copernicus published *De revolutionibus orbium coelestium*, we have known we weren't the centre of the universe. To call the aftermath of Copernicus' insight a mere 'revolution' feels profoundly inadequate, though the pun was too compelling to avoid, and perhaps impossible after Copernicus himself named his seminal work, in English, *On the Revolutions of the Heavenly Spheres*.

The idea that we are not at the celestial heart of the universe sometimes remains difficult to grasp, at least for those of us whose traditions were shaped by the God of Genesis decreeing that mankind shall 'be fruitful and multiply, and fill the earth and subdue it' and have 'dominion' over nature. But if, now, we accept that nature is more than a mere resource for us to exploit, then however we might describe our relationship with it – whether we are nature's stewards or live in mutuality with it or inseparably from it, or whether the relationship is so infused with mystery that it is beyond the domain of words – it seems to me there is an inescapable conclusion: we are failing creation.

We certainly failed the Brisbane River cod.

~

Once upon a time … How beguiling that phrase. How it transports us to lands of princesses and ogres, of good and evil plain for all to see. Once upon a time, Brisbane River cod swam in this river, while its closest genetic relative, the Mary River cod (*Maccullochella mariensis*), swam in the Mary River to the north. In some inconceivably distant past those two distinct species were one, and in turn had spawned from the Murray cod (*Maccullochella peelii*) on the western side of the Great Dividing Range. Geological time did its slow work, separating catchments that once were one, unconnected rivers now carving their own ways through their own river valleys. In the Mary River catchment to the north, the cod evolved separately to the cod in the adjacent catchment to the south.

Once upon a time, Brisbane River cod called this river home. How do we know? Because less than a hundred years ago we fished them, in enormous numbers, out of the river. Less than a hundred years ago this riverine habitat was healthy enough to nurture their young. In the 1930s, Brisbane River cod still swam in this water. The last rites were administered by bushfires that raged so furiously they blanketed the river with ash, suffocating the remaining fish. When we do our extinction work, we work quickly.

Extinction. It's part of our terrible genius as a species that we can know what we are doing to our fellow creatures and yet continue to stride forward – cruelly, callously, indifferently, helplessly – into an eternal tomorrow.

But we're also a hopeful species.

When I was here last year it was at a pool with a group of fish people from the valley, guardians of the river's fish stock. Garry was the leader and chief organiser of their bold, half-mad plan: to populate the river upstream of the dam wall with Mary River cod, the closest genetic relative of the Brisbane River cod. The 'pool' is a stretch of the river, slow and wide, shaded by cliff on the far bank, and overhanging branches where the river narrows and deepens. It's a permanent pool, big enough for adult cod to survive during dry times, and with enough protective shade for them to breed and rest during the day. Cod live most of their lives within a few kilometres of where they're born, spending their days under their favourite log. But to mate they'll travel three hundred kilometres.

Garry and one of the shire councillors – Jason, a towering man who is also a local veterinarian, and on whose riverside property I hope to camp on in a week or two – waded out into the river, standing in a patch of reeds with a bucket of fingerlings. Garry half-submerged the plastic pail into the river for a couple of minutes to allow the temperatures to even out, then passed it to Jason, who lowered it completely into the river water. I was too far away to see the first of the fingerlings leave the bucket, but Jason's and Garry's faces recorded the moment. Beaming smiles and wonder-filled shakes of the head. Applause and hooting from the audience of fishermen and fisherwomen.

'This is a selfless thing they're doing,' Jason said to me later, 'a community-minded thing, because they'll never see the fruits of their labour. These men will die before these fish reach maturity.'

If they reach maturity. Because waiting for them in the river are swarms of invasive and aggressive rivals, the introduced gambos (*Gambusia holbrooki*) and tilapia that have grown rampant in these pools. One of the men waded into the shallows with a net to show me how serious a problem they were. A single sweep produced a swarm of these noxious fingerlings.

'Evil bastards,' he muttered, tossing them onto the bank to die, flickering and glinting in the sun.

~

It's darkening. The gathering storm is sucking light and sound from the day. No bird, no beast but Dominic and me in the riverbed. There is nothing to say, and whatever we're walking into is bigger than us. The steel tips of our walking sticks crack against the river rocks. Who knows what power we may possess? Thunder begins to rumble around us. The wind and grass start to hiss. There's a clap of lightning so close I jump. So close it's almost like Dominic is aflame in that great jagged flash of light. He merges back into the dark afternoon more quickly than my heart returns to equilibrium. A father instinct suddenly hits me. I'm responsible for my son out here in this storm. Shouldn't we be taking precautions? But there's no shelter in sight, and the best I can think of is staying clear of lone gum trees, retracting our walking sticks so their steel tips aren't in contact with the ground and putting a little distance between us and the water.

We step up off the stony bed and take a cattle pad that's running parallel to the river. It's dark and it's loud and any moment now the universe is going to explode. It's exhilarating as hell.

'Isn't this great?' I yell.

Dominic grins.

Then it bursts. Lightning. Thunder. The sky a fury of rain arrows.

Dominic thinks he sees, through the slashing rain, a bridge. I raise my head to look but have to bow it again almost immediately as the wind whips the stinging rain into my eyes. A hundred steps later I shield my eyes with two hands and look up again. Maybe he's right. Another hundred steps, and then another.

By the time we reach the Kilcoy–Esk bridge, we're drenched. The sandy bed cascades in neat terraces, pylon to pylon, from the high bank on our right, towards the river's stream. We take shelter beneath the bridge, its concrete ribbing ten metres above us. We shrug off our packs and peel a Mars bar each. It's a quarter to three. On a clear day we'd have a couple of good hours of daylight left, but it's already grey and dark. Reckoning the time to our campsite just this side of Harlin village depends on what route we take.

Downstream of the bridge the entire fifty-metre width of the bed this side of the river is covered with head-high grass. Green panic. Surely there's a path through it somehow. I go down to the waterline. The rising water has covered any route there might once have been down there. Even so, I inch along, driving my walking sticks deep into the mud as third and fourth legs, leaning out over the water, peering downriver, hoping that a way may open just ahead. There's nothing. Just water and thick grass. I retreat. Perhaps the grass will thin somewhere in the middle of this field, perhaps there'll be animal tracks, or long clearings. Perhaps it's a mirage, or a comic book test, and if I walk at it confidently enough it will dissolve. I walk into the wall of grass. It does not dissolve. I plough forward. It's above head height, unrelenting. Again I retreat. Dominic is resting against his pack and finishing his second Mars bar as I re-emerge, despondent. He raises his eyebrows in cool query.

What else? I check the satellite map. We could cross the bridge to the other side of the river, follow it downstream through paddocks grazed low by cattle. But we'd need to cross back to this side of the river to reach our destination for the night, and the river is far too wide, too deep. There's an alternative – the nearby abandoned railway line. It means coming off the river, but it's the only option.

We make for the high bank, walking beneath the bridge as far as we can. Water pours through cracks in the deck, curtains of it, boring holes in the sand, driving onto our heads and shoulders and packs as we pass through. We step out from under the shelter of the bridge and scramble up the bank, over litter and through lantana. I turn to check Dominic's progress. He's right behind me. A large spider is making its way up his arm. He looks down, doesn't flinch. I brush it away. We reach the highway guardrail and wait until the bridge and its approaches are clear of traffic before levering ourselves up and over. For a moment we stand on the bitumen in the steady rain, panting, Dominic bleeding where he's just cut his palm on an errant shard of steel. But then a truck appears and begins to bear down on us, and we scurry off the bridge before it reaches us.

On the other side of the highway heading into Harlin we can see the old railway, the same one we'd encountered at Emu Creek. We make for it, scaling a gate, crossing a paddock and climbing a boundary fence before stepping down onto the rail trail. The sleepers and rails have been pulled up and the ballast scooped away. This narrow path is now the domain of cyclists and horseriders. The ground is beautifully even, and the trail stretches straight as far as we can see. After all the days of rugged and undulating riverbank terrain, this should be easy. We stride out, the final two kilometres of the day ahead of us, the rain beginning to ease a little. Flat though it is, the ground is compacted and hard. We're fenced in either side. Crunch, crunch, crunch. After the life of the riverbed, there's a desolate monotony here, and suddenly, after seven days of walking, my feet are sore. My knees creak, and my left ankle begins to ache.

~

We're completely soaked when we reach our campsite. Or rather, reach the front gate of the property owned by Catharina and Steve, who've agreed to let us camp at their place tonight. Catharina comes out to greet us. Our packs are wet despite their covers, our clothes are soaking and our boots are waterlogged. She leads us to the shelter of her porch with its view of the river. We look through the rain and the early dusk. Three distinct channels are forming in the rising river. Somewhere down there, through the steadily falling rain, somewhere on that slippery bank, is the spot I'd arranged for Dominic and me to pitch our tent.

Now, with our packs off our backs and our bodies cooling quickly after the day's exertion, I shiver.

'We really would be happy if you slept up here,' Catharina says. I've already accepted her offer of dinner.

'We've got a room each for you. You can dry your clothes and have a hot shower.'

The words 'hot shower' are irresistible to a shivering hiker faced with pitching a tent in the dark and rain on sodden ground. Plus, I tell myself, I have my son to think about.

At the dinner table we hold hands for Steve to lead grace. Steve is on my right, Dominic my left. I like the contact, the warmth. I like the small circle we create. I like being with these two kind people. 'Amen,' I say, and thank them once again for their hospitality. At the entrance to their property is a large sign, *Harlin Garden*, with paintings of squash, broccoli, turnips, carrots, pumpkins and apples arrayed above a golden dawning sun. Catharina serves the vegetables this evening, a similar smorgasbord, including – though it doesn't feature in the spray-painted still life – Dominic's first choko.

They want to know about how difficult the terrain along the river is.

'Don't you ever feel like taking the road?' Catharina asks.

The question that arose in the first days of the river as we crossed the thirty-eight causeways over the Western Branch, each crossing a temptation to take a short cut. I shake my head.

'It's as if the river is talking to you,' I explain. 'It wants you to follow it. "This way," it says, "this way," and so you accept its invitation.'

Catharina smiles. 'Well, the river says it's fourteen hours from Linville to here.'

For the floodwater that is, when the river first begins to rise after rain. So we talk about floods. She leads us outside and points to where the water got to in February, and where in 2011; it was lower this time around, but not that much. She shows us video footage of three of Steve's cows trapped on what became, in the floodwater, an island in the river. They swim towards the camera. They swim over the submerged boundary fence, into Catharina and Steve's backyard. They swim to safety.

~

The rain begins again at 4.20 am, heavy on the roof, heavy on my hopes for the day. I lie awake. The bed is almost too soft. Do I admit to myself that my feet are sore? At this time of the morning, there's no holding back these little anxieties. Large blisters, yes, on each heel and each big toe, but that's not what worries me. As long as a blister doesn't get infected, you can walk through it. No, it's the sides of both heels, especially my right. They're aching as if bone-bruised. I wince as I roll out of bed and get up to check our drying boots. The tiled floor is cold and hard and brutal on my bare feet. I reposition the boots before the pedestal fan in the laundry and pad back to bed, relieved to take the weight off my heels. While my legs have become stronger with each passing day, my feet are more tender. After half-an-hour the rain ceases and I drift back to sleep.

Day 8

Harlin to Scrub Creek

Distance: 17.38 km

Evening Camp: Place of the stone causeway

It is Sunday. The shape of the week has disappeared into the journey, but I know it's Sunday because Catharina and Steve must leave at 8 am to drive to church, an hour by car over the D'Aguilar Range east of the river. We leave then too, our wet-weather gear on, and take the rail trail for the last kilometre into Harlin.

The village is still asleep. The lowset, brick Harlin Hotel is painted in the colours of parochial Brisbane's beer, bright gold, embellished at regular intervals with four red Xs. The primary school is deserted. A corrugated iron shed outside the school serves as a bus-shelter, a red Australia Post letterbox its companion. There are no vehicles filling up at the petrol station, empty at this hour. A row of silky oaks line the road. Relics neatly decorate the abandoned railway siding and an enormous lemon tree grows near where the stationmaster's house might have been. We ease through the village and cross Ivory Creek on the outskirts and follow it, past the remnant of the railway bridge destroyed in the flood of 2013, down to the river.

Our destination today is Scrub Creek, seventeen kilometres away.

~

The sky is close and grey, the rain soft, more mist than rain. We stride out. It's good to be down in the river, the two of us and all this gently falling, rising, flowing water. We're moving through cloud. Puffs of mist rise from the river up over the hills like smoke. Pale-headed rosellas make their own way above us. In the fringe of trees above the high bank, crows wrestle branches to themselves.

And then I see the first pelican of the trip. The thrill is acute.

I turn to Dominic, pointing – 'Look, a pelican!' – to make sure he sees it too. It's almost aglow in the mist, its white breast, its long neck, its pink translucent beak, as if some sacred light has fallen on it. I know I'm smiling. Is it possible to smile too readily? To fall into holy foolishness?

Life, depending on how you look at it, is a continuation of firsts. The first pelican of the trip. The first pelican we've seen today. The first we've encountered in the last hour. The first on this stretch of river. Perhaps the trick is staying open to magic.

The pelican outpaces us; down it goes, leading the way. So where are you taking us, old bird?

~

On the left bank, elevated above the river, is a quarry. It is unmissable. I was expecting it but still wasn't ready. The hillside is scarred by the cleared vegetation and exposed quarry wall. Igneous andesite porphyry is mined up there. It's good rock: hard and durable. Perfect for road base, concrete aggregate or railway ballast.

On the right bank, low, off-river, opposite the hard rock quarry, is a second quarry, its sibling, this one an alluvial plant. Because rivers aren't just a source of water and food. The alluvial bed of a river like this one is composed of material humans have used for millennia – sand and gravel. From our vantage point down on the riverbed, the second quarry is nearly invisible, only the cabs of a fleet of trucks rising above the line of the high bank.

It's Sunday and both sites – left and right – are dormant. The caretaker spots us and drives down from the plant with his young son in the passenger seat of his truck, both wearing quarry caps. We meet on the private causeway linking the two quarries. I introduce Dominic and myself. I'd spoken with the quarry owner previously about the walk, and the caretaker is content.

The river is both parent and midwife to its sand and gravel quarries. Eons of flood and flow have scoured and abraded and gnawed and chafed and ultimately deposited the sand and gravel into their current resting place. But not only has the river been ageless creator of the gravel, it is also the source of the water needed to mine and process the quarried material. Its creative work is redirected into extractive work.

Six months earlier, a court had denied another hard rock quarry and concrete batching plant's application to set up an operation a couple of kilometres downriver. That grazing property had once upon a time quarried its alluvial material but ceased some decades ago. 'Save for changes to topography in the area where sand and gravel was extracted, there is no other evidence of this historical use of the land,' the court decision reads, as if changing the very topography of place is a thing of little moment. The quarry would serve the local community, the applicant argued. And if a community needs rock, surely it's not a bad thing for that rock to come from the local geography, for a community's buildings to rise from its own bedrock. Sure the quarry might not be attractive, but neither was it unduly ugly. Could the site have been rehabilitated? The judge thought so. And while the proposed quarry would need a lot of water – some recycled or harvested, but most extracted from the river – and didn't yet have the proper licences, there was no suggestion it wouldn't, eventually, be given permission to have all the water it wanted. But cutting earth from itself cannot be done silently, and this quarry would have been too noisy. So the judge determined in rejecting the owner's application. The property downriver remains a cattle station.

At the causeway, the caretaker wishes us well and sets us on our way. In ten years' time it may be him and his son walking this river. We cross over and take the inside bank as the river sweeps left. We pass a blue pump on steel skis drawing water from the river, the skis making it easier to drag the pump up the bank in flood. The grey sky lowers itself over us, and we begin making our way through patches of intermittent rain.

~

Downriver we walk, always downriver. There's a creek ahead, always another creek. This one – Gregor's Creek – is too wide and deep to cross at its junction. We head upstream, the banks steepening the further we leave the river behind. We shimmy across a fallen log, then drag ourselves up the bank. There's little purchase for our boots, so we lever ourselves forward with our walking sticks, inch by inch, setting our route between gum trees, pulling ourselves up, resting against the trunk, allowing it to hold our weight while we catch our breath, before setting out for the next gum. The incline eases off and we are near the top and in a grassy paddock and there is a barbed wire fence yet to navigate, and then behind it a bitumen road is dropping back down towards the river and the Alf Williams Bridge.

Down on the stony pan, there's movement by my right boot. I stop. A tiny toadlet disappears beneath a rock. I squat and lift the stone, and away it leaps, ricocheting off a nearby rock, and then over that one, and away. It's a cane toad: invasive, marauding, poisonous. I stand upright, and as I straighten I see another. And then another. My field of vision steadies and I realise the pan is alive with thousands of toadlets. A plague of them. It's as if the Lord God of Exodus has turned his wrathful attention from the Nile and Moses' Egyptian captors and sent a biblical plague of toads our way. 'The river will teem with toads. They will come up into your palace and onto your bed, into the houses of your officials and on your people, and into your ovens and kneading troughs.' But if this plague is retribution for a sacrilege, who is the hard-hearted pharaoh, and what divine message are we refusing to heed?

On the high bank to our right, two white cedars – yellow-leaved, deciduous – glow in the muted light. A whipbird calls from across the river. Beyond it a high hill is crested by a modest stand of hoop pine within a conservation park. That question again: what did it once look like? Like that, I think, like that.

~

The river sweeps to the right. We sweep right, on the inside. The river straightens. Ahead a dark blue ute pulls down off the high bank and into the bed, then disappears. We navigate a patch of callistemon, and when we're through see the ute again, a couple of hundred metres ahead, stationary in the bed, blocking our way.

The engine is running and the man in the passenger seat, nearer us, has his window down. They're clearly waiting for us at what we can now make out is a private causeway. They're young, jackaroos by the looks of them, the name of their cattle property stitched into the breast pockets of their workshirts.

They glare at us.

'Hey there,' I greet them, smiling, my thumbs hooked into the straps of my pack.

But they won't be distracted.

'What are you doing?'

The question is loaded with accusation.

I introduce myself, first and last name, and then Dominic, tell

93

him we're father and son and that we're walking along the river. My tone is light. But his eyes are narrow.

'It's private property.'

Private property, as if nothing more need be said. Private property, the beginning and the end of all possible argument. The source of all authority, all worldly power.

'I've spoken with the owner,' I say.

He lifts a mobile phone to his right ear, and I realise he's had it on all along.

'He says he's spoken to the owner,' the jackaroo reports to whoever is on the line.

'I've spoken to over a hundred farmers along the river here,' I continue, 'and I'm sure I would have spoken to the owner.'

'He says he's spoken to over a hundred farmers.'

'We started up the top of the Western Branch a week ago.'

'He reckons they've been walking for a week. He reckons they started up the top of the Western Branch.'

'I've spoken to over a hundred farmers,' I say again. 'I might have missed someone – I guess it's possible.'

It's becoming obvious that the jackeroos' job is to hold us until the station manager gets here. They've caught a couple of troublemakers but have to wait for their boss to decide what to do with us.

I glance at Dominic. There's a hint of a smile at the corner of his mouth, and I know what he's thinking: So, Dad, how are you going to talk your way out of this?

Eventually a white HiLux appears on the high bank of the other side of the river, making its way towards the causeway, and us. The jackeroos give nothing away; their faces are inscrutable. I step back from the ute. We fall into silence. The rising river is flowing over the causeway. The HiLux drops down the bank and moves steadily across, parting the water as it comes.

I raise my arm towards the HiLux, palm open in greeting, and take a step towards where the property manager will pull up. He is in his late thirties and has an open face. He hears me out, and looks at Dominic beside me. He sees a wet fifty-year-old bloke in hiking gear who's more or less who he says he is. And he's curious. When did we start? How many kilometres? Where are we aiming for tonight? Nodding at every answer. He explains whose property this is, both sides, most of the way down to Scrub Creek on the left. Says it'll be easier walking on the left bank, says it's fine for us to go that way.

'But watch the river. There's a fresh coming through.'

We thank him. He pulls away from us to turn his truck around. Dominic and I walk down to the water's edge. The river is loud as it courses over the concrete ford, beginning to drag sand and quartz across the causeway as it goes. We pause to weigh up what to do. Our boots aren't yet waterlogged and we figure it's best to keep them as dry as possible for as long as possible. As we look for boulders to sit on while we take off our boots and socks, the station manager pulls up beside us on his way back.

'Get in. I'll give you a lift across.'

We throw our packs in the tray and climb up. I steal a glance back at the jackeroos in their ute. Am I imagining it, or do they look disappointed?

~

Private property.

The law can sometimes be a thing of beauty, a shield to protect and a sword to liberate. But the laws that regulate private property – at least in the legal traditions of Britain and continental Europe – are as complex as they are prosaic. Libraries and lifetimes dedicated to legislating, codifying and clarifying property rights. We've done that in pursuit of certainty, an elusive objective, to make possible the accumulation and protection of wealth, most often private wealth.

But if you zoom out far enough, the idea of private ownership of land grows absurd. We might carve up continents into private parcels of geographical space. But entire planets? Solar systems? The boundless universe? Or zoom in close and it's no less strange. I might own the land upon which I build my house and plant my garden, might own that topsoil. But under a microscope I realise that each handful of dirt is a universe of life invisible to the human eye, a teeming world of micro-organisms. Do I own those wild creatures too?

Once, not very long ago at all, Jinibara and Jagera and Turrbal law would have guided who could walk along the banks of the river. Where. When. Under what conditions. Back then, before the British arrived in the valley in the 1820s, the idea that the law of a colonising power such as those spawned by Europe might supplant that of the owners of this river country would probably have been inconceivable.

I think of the series of five letters published in the *Moreton Bay Courier* in 1858 and 1859, addressed to the editor, each signed off 'DALINKUA, DALIPIE, Delegates for all blackfellows. Camp, Breakfast Creek.' Breakfast Creek, a downstream tributary of this river as it flows through the city, had – in the little more than thirty years since whitefellows arrived in the early 1820s – witnessed much. 'Hear us now in the indictment we bring against our Anglo-Saxon brothers at the bar of Universal Justice,' one letter begins. 'These, our white brothers, have taken our hunting and fishing grounds from us ...' The arguments prosecuted by Dalinkua and Dalipie are sharp, though less fierce than they had a right to be. The letter continues, in part:

Now how did our pale-faced brother obtain our land? ... he merely
ran it over with a theodolite, a chain, a hatchet and some pegs or

some marks on trees – their titles patent from the Queen, it thus became his property ... Sometimes we are driven away from any river frontage, where we could fish and find game ... On those who assume the ownership of our land, and grant leases to occupy the same, making no provision that we shall be supplied with food, must and does lie a heavy responsibility. ... We were here before white men, and while the habits of our white brother adapt him to live in any climate, no part of the globe would suit us to live on but this the country of our birth.

So what now? To what extent do the laws of theodolites and land-clearing hatchets and pegs in the ground and grants from the Crown still govern us? We live in this most fleeting of moments in the smallest of slivers of an inconceivably great sweep of time and history, and we rely, still, on something as crude and complex and brutal as the law in our attempts to regulate our affairs.

For this river, at this moment, the law works something like this.

First the law has to define what a river is. Is it just the flowing water? Or does it also comprise its bed and its banks? And if so, what does that mean for a river as sinuous and shifting as this one; whose natural cycles encompass both droughts – when it will shrink to a series of pools – and floods?

The law the colonists first brought from Britain adopted the Latin principle of *ad medium filum aquae*, literally, 'to the centre thread of the water'. In practice, a landowner with river frontage owned to the midpoint of the river: half the river was theirs.

But then the parliamentarians with their statutes stepped in to swap the simplicity of *ad medium filum aquae* for the complexities of legislation. Now, in Queensland, water and land are treated differently, each having its own statute: the *Water Act 2000* and the *Land Act 1994*.

The *Land Act 1994* divides the world into tidal and non-tidal watercourses. Then it creates the notion of a 'boundary' between someone's land and a non-tidal watercourse that 'adjoins' their land. (A committed reader of the statute must go to another statute – the *Survey and Mapping Infrastructure Act 2003* – for guidance about what a 'boundary' is.) And it tells us that the land that 'is on the watercourse side of the boundary is the property of the State'. Such land is also known, in countries that adopted the law of England – like Australia – as 'Crown land'. An earlier version of the legislation put it more succinctly: the Crown owned the 'bed and banks' of a river. Which means that the state owns the sand and the gravel and any other quarriable material residing in those bed and banks.

Though the state owns a river's bed and banks, landowners have some rights over what goes on in the bed of a river. But what rights exactly? If the owner has a right to take water from the watercourse (whether or not an owner does is regulated by the *Water Act 2000*), then they also have limited rights over the bed and banks: such as the right to access that land and to graze stock on it. So the law protects graziers. It is true, as Graham said at the start of this walk, that this is cattle country. But the law gives owners another right too: the right to evict others off that state-owned land – to 'bring action against a person who trespasses on the adjacent area as if the owner were the registered owner of the adjacent area'.

So no, it was not private property where Dominic and I stood in the bed of the river, between its high banks. It was land owned by the state. But yes, the owner of the adjoining land could ask us to leave.

~

And so we approach Scrub Creek, expecting a rough stream struggling through unprepossessing bush. But the causeway at Scrub Creek is a thing of beauty. It's much older than the thirty-eight we'd crossed a week or more ago on the Western Branch, which could claim, at best, that they'd survived a flood or two. On the downstream side of this causeway a fleet of stone terraces cascades down to the water level, the stone locally hewn, irregularly cut, each block unique. Even its more recent concrete is stained the colour of the damp sand that was heaped along the left bank by February's flood. Under a stone-grey sky, the causeway appears to have grown from the landscape itself.

~

Our stop for the night is the first property downstream from the causeway, on the right bank, a vegetable farm run by Sharelle and Paul. Sharelle is a daughter of the river, now a lawyer. Paul is forty but looks younger, also river born.

'Two years ago I didn't know the first thing about onions and broccoli and pumpkin,' he'd told me when we first met. He'd been running cattle and deer for years, and will again. For now, he's growing produce, learning swiftly, working long hours, bringing his lifetime of living on the river to this new type of farming. 'This alluvial loam is extremely fertile. During the 2011 flood this property picked up a foot of topsoil. You wouldn't believe how good the yield here is. Nearly twice as high as in the Lockyer.' The Lockyer Valley, named after the creek that flows over a hundred kilometres before joining the river just below the dam wall. If all goes well, I'll reach the junction of the Brisbane and the Lockyer in six days.

What a good farmer can do with their land is of urgent importance. But it's not, I have to admit as I arrive at Sharelle and Paul's farm, the produce that has me excited. When I'd visited Paul some months back, he'd taken me down to a deep hole in the river full of lungfish – a dozen or so, he told me. I couldn't tell it was a hole. It looked like a stretch of river, but Paul knows that if the river dries up, it's holes like this that remain.

'Lungfish? Really?'

'You can see them at sunset,' he'd said. 'You can hear them during the night.'

We know humans have been on this continent for 60,000 years or so – the proverbial blink of a lungfish's 380-million-year-old eye – whereas *Neoceratodus forsteri* was here before dinosaurs. These creatures are thought to be our closest living fish relatives. These fish in Paul's hole closely resemble fossils of lungfish that lived 100 million years ago, and evolutionary biologists believe that relatives of these creatures evolved adaptations – lungs, fins of bone and muscle a little like human forearms – to pull them out of the water and up onto land and ultimately to become walking, air-breathing vertebrates, like humans.

Paul showed me footage he'd taken on his phone during the last drought. There he is delicately scooping the great fish out of the shrinking pool in the river with the bucket of his tractor so he can move them to a larger pool, to safety, survival. One of his daughters stands beside the tractor, barefoot, reaching out to the giant squirming fish with her hand.

Paul and his daughter and the fish they were so committed to protecting set me dreaming. I imagined lying in my sleeping bag at night beside this hole in the river and hearing a sound I'd never before encountered, as the mythic fish rose to the surface to breathe. Or better still, pulled itself out of the water and dragged its great body along the mud towards our tent, this fish with both lung and gills, this prehistoric creature, the link between fish and amphibians. Among the things I really wanted to do on this walk, sleeping on the banks of this lungfish hole was one of them.

~

It starts raining steadily again. We follow the road as it rises up out of the river towards Paul and Sharelle's farmhouse. An eastern brown has been crushed into the bitumen by a recent vehicle.

A young girl, the one with the lungfish in the video, still barefoot all these years later, picks us up at the farm's front gate on a four-wheeler. She can only be ten or eleven years old, but she's serious and in control, and moulded from the land. We leave our packs under an awning at the back door and Sharelle joins us for a tour of the farm. Peacocks and chooks and roosters and three Jack Russells tear after us in the rain and mud.

'What crops have you got in?' I ask.

'Those,' Sharelle points, 'are potatoes we planted after the February flood. We were lucky. If we'd planted any earlier we'd have lost the entire crop.'

'But those pumpkins over there,' she says, her hand sweeping across another field, 'we're going to have to plough back into the ground.'

Because the price for these pumpkins is currently so low it'll cost more in labour and fuel to harvest them than what they'd get at market.

And so, in adjoining paddocks, are the counterweights of farming's fates.

~

We humans have arranged our affairs in such a way that we will plant a crop, spray it, watch it grow, pray it won't be drowned in floodwaters or shredded by hail or consumed by locust, and then, when the season for reaping arrives, stand aside to let the crop die. It's troubling that men with as fine a feel for growing crops from the earth as Paul, as full of care and responsibility for the land, have no option but to turn their backs as their efforts come to nothing. There is, it seems to me, a wrench to the natural order of things, a tear in the soul of our communal being. But what is to be done? The market is master. And perhaps the market is no less cruel and wilful and unfathomable than nature is.

But think of the water wasted. Think of how much water those pumpkins – which are mainly water – needed to grow. They take four to eight megalitres of water per hectare. But those numbers evade meaning. When it comes to water systems, I've come to see the world in glasses, Olympic swimming pools and Sydney Harbours. They are tangible units, the modern equivalents of the horsepower and hands and feet that people once used to navigate farm and village life in England. Those old measurements were rooted in a lived experience before the industrial revolution, let alone the digital one. It's not just that human systems are now bewilderingly more complex. It's that the scale and volume of our interactions with the world around us has changed. New colloquial units of measurement are a bridge between science and salvation, essential in navigating modernity at our new human scale. We still need to know the natural world through our blood as much as our minds if we are to nourish it back to life.

So, a crop of pumpkins takes four to eight megalitres per hectare. I figure this paddock is about a hectare. I grapple with the maths. If an Olympic pool holds about 2.5 million litres (fifty metres long, twenty-five metres wide, two metres deep), or 2.5 megalitres, then the paddock I'm driving beside in a four-wheeler has needed a couple of Olympic swimming pools full of water to grow those pumpkins. Two Olympic pools of water, wasted.

But, I then think, is 'wasted' quite right? Because you cannot destroy water.

~

'What chance of seeing lungfish in this weather?' I ask Paul.

'Nah, mate.' He shakes his head slowly.

And so the rain puts paid to my dream of camping out with lungfish.

Sharelle offers us the downstairs floor of the farmhouse. Dominic and I hang our clothes to dry, and after a post-dinner round of spoons with Sharelle and the kids, we roll out our swags. Dominic will be leaving tomorrow, and my brother Ian arriving. I flick the light switch off. Still the world turns. The peacock screams from its cage outside as the male Jack Russell pads across the tiled floor to where we're lying for a final curious sniff. Down in the waterway, lungfish rest in the dark, the river's skin atremble above them.

Day 9

Scrub Creek to Mount Beppo

Distance: 18.2 km

Evening Camp: Place of the flooded lagoon

The house wakes before dawn. Roosters and rain, farm kids to be readied for school, a lawyer with a week of clients ahead of her, a farmer aiming to use the wet for a day's welding in his shed. It's a formidable domestic operation, and each member of the family knows their role. Daylight, when it arrives, is subdued by heavy cloud, and our farewells, when the time comes around, are muted.

Dominic and I are alone in the house now, waiting for my brother Ian. Our boots are as wet this morning as they were last night. My sister Kate is driving up from the city with Ian and will take Dominic home. She's also bringing fresh supplies. I busy myself assessing what I'll need for the next seven days, tallying what I haven't used, and therefore how much I'll need from Kate. When she arrives I'm thrilled she's also brought one of our nephews as company. We all embrace warmly in the lightly falling rain.

I don't yet have the routines of resupply down pat, but I am quicker this second time, surer in what I'll need for the next leg. The others talk while I kneel on the floor, filling my backpack.

I hear Kate ask Dominic what it's been like and hear him struggle to answer.

Kate must get back to the city. I walk Dominic to the car, my arm round his shoulder, his around mine.

'See you, Simo,' he says.

'See you, Domenico.'

We let each other go but will, I'm sure, hold close our three precious days together.

So I farewell my son. But welcome my brother.

Ian is five years younger than me, a little stockier, hazel-eyed to my blue. There are a couple of centimetres between us. He is bearded as I, on this walk, am becoming. We both have two sons, though his are younger. Ian went to university to study plants and then spent a decade working in national parks. He lives simply, sparely if he must, and is a great walker. He moves easily through the bush and carries his cares more lightly than I do. In a family of five siblings, this will be the first time we've ever spent more than a day together, just the two of us.

But before setting out there's a call I need to make. I need to speak with one of the senior members of the first settler families about the route ahead, where it passes in front of his property. The river is up, he says, so it's not possible to walk along the bed. He's sceptical of our chances of getting far today, or indeed getting through at all. On his bank the grass is chest high, and he's been seeing snakes every day. He'll give us his blessing if we insist, but not his encouragement. I thank him – both for his generosity, and no doubt his good sense.

~

The grass *is* chest high. We can't afford to think of snakes, so we wade through. Chest high, and thick, and at some point that refrain returns: What was it like? Did there use to be grass here? Did the grass ever grow this high?

There's a diary of one of the white explorers, Edmund Lockyer, who travelled up the river in September 1825. Further downstream near the banks of the creek that carries his name, Lockyer had observed that the traditional owners 'had lately set fire to the long grass and the new grass was just above ground, making this plain appear like a bowling green'. But further upriver on that expedition, about where we are now, it was a different story. 'Here the bed of the river, though broad, was nearly dry … the long and thick grass made it dreadfully fatiguing to walk through with our loads, the men being badly off for shoes, and two of them having sore feet.'

So, grass will be grass. And wading through the really tall stuff on foot has always been exhausting.

~

What was it like? The traditional owners of this country managed it, set fire to the long grass, drank from the river, fished it, called it home. They managed it for a very, very long time. What happened? I have no capacity to arrange words to tell the stories of their dispersal from the river's waterways and flats, no capacity to do those stories justice. I find myself floundering before the scale of what occurred, its import. The dispersal was too swift, the work of a short few decades, from 1841 when the first colonists swarmed up the river valley, or entered it over the D'Aguilar Range from the east. That history can, at times, do its remorseless work as swiftly as it does is shocking. That the identity of those of us who usurped land or pulled triggers or poisoned flour might be concealed behind words like 'history' is shocking.

But one story from this catchment recurs again and again in conversation and in silence, as if whispered by the river itself at every bend.

The squatters who arrived in the river valley seeking their fortune in 1841 were the sons of aristocracy. The McConnells claimed Cressbrook, the Balfours (maternal uncles of Robert Louis Stevenson, who was christened with the 'Balfour' but dropped it at the same time he changed the 'Lewis' of his birth to 'Louis') claimed Colinton, the Bigge boys ('Big' Bigge and 'Little' Bigge) claimed land at the foot of Mount Brisbane, and the Archer brothers from Norway claimed country then known as 'Durundur' in the Stanley catchment and retained the name. The Archers' neighbours and fellow-squatters were the brothers Mackenzie, who took what they called Kilcoy station as their own.

And so, in late 1841 or early 1842, the Kilcoy massacre occurred. There are no eyewitness accounts, no photographs or frank journal entries or deathbed confessions. But what news channels there were ran hot. Soon after the Commissioner of Crown Lands, Dr Stephen Simpson, was appointed at Moreton Bay in May 1842, escaped convicts brought the massacre to his attention. At least thirty First Nations people had died from the effects of food given to them on a station. The Mackenzie brothers' shepherd was killed in revenge three months later. Soon enough news reached Sydney and, by 5 December 1842, the pages of the *Sydney Morning Herald* reported the words of a German missionary passing through the valley: 'A large number of natives, about fifty or sixty, [had] been poisoned at one of the squatters' stations.' Simpson himself travelled through the district in March and April the following year, and continued to hear stories of the killings from the local Aboriginal people he met. Station staff would say only that of course there had been arsenic on the station, and that as everyone knows it was for treating scab on sheep, and that sometimes sheep died from ingesting it. Perhaps some Aboriginal people had died from eating the carcasses of poisoned sheep. Yeah, right. That wouldn't be the evidence of Captain Coley, who appeared at the May 1861 hearings into the Native Mounted Police Force, and had known Evan and Colin Mackenzie. Coley was clear about responsibility:

the Mackenzie brothers had been involved in poisoning Aboriginal people on their station by putting arsenic in flour.

And so the 1840s in the catchment of the river and its tributaries was a decade of conflict. Stack the deaths, one upon another. Killing after killing. A decade of guerrilla warfare. A decade of terror.

~

Over to the right, set well back from the river, is the old homestead complex, Cressbrook. The run, taken up – as the old parlance went – in 1841, was the first outside the fifty-mile settlement limit of the Moreton Bay penal colony downriver, nearer the river's mouth. Amazingly, that first settler family is still there, red stag on its family shield from when it accepted responsibility for those first deer imported into the colony.

We leave the old family's property behind and pull down off the high bank, looking for respite from the interminable grass. A spew of ugly 'Mother of Millions' weed greets us, followed, a moment later, by a pair of brilliant scaly-breasted lorikeets in an old blue gum. Both are firsts on this river expedition, this yo-yoing of the heart, despondent sigh followed by delirious song. Ahead is a frog-loud marsh, impossible to cross, forcing us back up the bank, all the way to the high bank and a barbed wire fence and an unexpected crop behind it.

It's not, is it? I'm first to crest the bank.

'Is that …?'

But the possibility seems too preposterous to entertain. Not here, surely.

'Cotton,' Ian says, arriving beside me.

Cotton had been ubiquitous in my childhood on the Darling Downs. Friends' parents grew it. Teenage school kids from Toowoomba spent summer holidays stick-picking in fields to prepare fields for planting. You'd see balls of cotton on the verges of highways, teased out of bales on the backs of hundred-kilometre-an-hour road trains taking the cotton to market. But cotton was a thirsty crop. On any drive out west you'd witness huge earthen-walled ring-tank dams being built to capture local run-off, interrupting the natural flow of creeks and gullies and rivers. You'd know about the cotton station as big as a country just over the border in New South Wales that denied the entire Murray-Darling its health. You knew how avaricious cotton was when it came to water, and how ill-suited it was to this dry continent.

But to see a field here, in cattle country, in a district of small holdings growing produce, beside a river that already has demands enough on it!

'Cotton,' I finally accept.

How tangled the thread of slavery and commerce that spooled across the oceans from the United States to Britain to this river valley in the nineteenth century.

On 4 March 1858, a slave-owning, cotton-planting South Carolinian senator stood in the American senate and proclaimed: 'You dare not make war upon cotton! No power on earth dares make war upon it. Cotton is king.' Senator James Henry Hammond protested too much. Change was on its way. Not that the senator would have been aware of Queen Victoria signing letters patent on 6 June 1859 to allow the colony of Queensland self-governance. Nor that the new colonial government of Queensland's first Act of Parliament, passed on 17 September 1860, was its *The Alienation of Crown Lands Act 1860,* which created incentives for settlers to grow cotton. By 19 April 1861 an American Civil War president issued his *Proclamation of Blockade Against Southern Ports* and began to strangle the revenue stream of his southern slave- and cotton-trading states. Across the Atlantic, cotton mill owners in Manchester went cold, their supplies cut off overnight by a presidential gamble. Cruelly, the abolitionist trade unions and their workers began to suffer.

Commerce has always been cosmopolitan. Profit follows opportunity. War upturns everything except mercantile ambition. And an Indian Ocean away from Britain, Queensland's colonialists were ambitious.

Three days after Lincoln's proclamation, on 22 April 1861, Scottish pastor and colonist John Dunmore Lang wrote his *Queensland Australia: A Highly Eligible Field for Immigration and the Future Cotton-Field of Great Britain*. 'It was a natural and warrantable conclusion that cotton could be grown by white men at Moreton Bay as well as by slaves and negroes elsewhere ... As cotton was an article of agricultural produce for which there is a constant and unlimited demand in Europe, while it is supplied at present to the European market almost exclusively by the labour of slaves, I could not help recognising what I saw as a certain prospect of adequate remuneration for the Australian farmer, should he betake himself to cotton cultivation.'

And so, in the 1860s, with America's economy in ruins and its young soul in torment, and Manchester's workers increasingly desperate, cotton plantations started to take a precarious hold on the banks of this river.

~

But flowers are innocent of history, and today this cotton crop is in glorious flower. The dark green field is tossed through with yellow and pink blossoms, delicate pastels under the day's grey mantle.

A cotton bud, when it opens, blooms in the morning, a creamy or yellow colour. By the following day it has pollinated, become pink, and is already deepening, turning. This field of colour before us is coursing with life, is coming into being. I lean over the fence and pull a branch towards me, a flower, with its tight fold of darkening pink petals, the outermost beginning, already, to peel away. On the ground I spot a cracked boll, the cotton already spilling out. Nestled in the fluffy white fibre will be its dark seeds.

Ian pockets the boll. We resume our journey.

Between the high bank and the water is a long terrace, planted out with hundreds of trees. Stakes and low olive-coloured triangular corflute protective barriers shield and support ti-trees and she-oaks aiming to hold the terrace, to prevent the bank from eroding and, I think to protect the crop in the paddock above us.

Up on the high bank again, the grass is thick and tall and conceals a barbed wire fence. Ian walks straight into it. He grunts and extracts himself. He doesn't want to look. Turns his eyes to the sky first, then to me. But there's no avoiding it. There's an ugly gash on his shin, pouring blood. He grimaces and shakes his head. He can't quite believe it. This is the second hour of a planned seven-day walk and he might already be a casualty. I pull out my flask of Jameson's whiskey; this is its first application. Ian winces. But the sting of whiskey on an open wound is a good sting, and I laugh. 'More,' he says.

At midday the river opens up, and we drop down into the bed. We cross the river at a riffle, and track down the left bank for a while. Ian explains the difference between a great egret and an intermediate egret – size, yes, but the larger bird also has a more pronounced kink in its long white neck – and stops to show me a foraging pipit in a patch of low grass. We begin walking into a breeze. A march of power poles runs parallel to the river on the opposite bank for a while before disappearing again, its wires gently dipping, rising and dipping once again. We cross back to the right bank at the next opportunity, not wanting to get caught on the wrong side when we run out of light.

~

We stop when the water channel is ten or so metres below us and curving gradually to the right. We take care not to stand too close to the edge. The bank is falling into the river, eroded from the last flood in February, and the one before that, and probably the one before that too. Sometimes the changing path of the river is only discernible from a vantage point of a thousand years. This one is being re-sculpted before our eyes. Low clouds are closing in all around us now, horizon to horizon, however much rain is in them. Perhaps the land we're standing on will, tomorrow or next week or next month, collapse into the water, and the river's channel will have altered just a little more.

A grey wind tugs at our collars and slaps against the rain covers over our packs. The grass begins to dance, wildly. We breathe in the changing air, looking around, centring ourselves. Ian sees it before I do and points inland across the flat paddock. His finger traces a long embankment running away from the river, across the paddock to our right, cutting across this inside bend. Soon enough I see it for what it is – a previous bank of the river – and realise the ground we're standing on now is merely the accumulation of tens of thousands of years of flood deposit, the silt and sand being carried here and there by the waters as the river meanders its way through time.

Meander. A beautiful, fertile, word.

The Büyük Menderes River of Türkiye, the famed Maeander of antiquity. That ancient river's god was also Maeander, son of Oceanus, who is carved into the Trevi Fountain in Rome. Rivers once had their gods. Perhaps they will again.

A river will do what it will no matter its name. Create and destroy. Bring plenty and inflict pain. Yet its name is also important. It tells us something of ourselves. How might we change, I wonder, if we called the river by a different name? Might sacramentalising a river protect it?

~

We reach a great sandy beach high up on the outside sweep of the river, eight or nine metres above the surface of the water. The beach is fifty metres wide and runs for half-a-kilometre. The sand is pure, the grain fine, deposited all the way up here in the last flood. It feels surreal to be standing on a sandy shore separated from its waters like this one has been. The golden expanse of sand elevated above the river as if by magic, now suspended between the river below and a flat grey sky we can almost touch.

There are no prints of human feet on this sandy shore. Cattle, wild dogs, and on its fringes, kangaroo. The rain, when it comes, will wash us all away.

~

The river bends to the left up ahead. I check my GPS. Yes, that's where we're aiming for tonight. It's been a long day, the longest, by distance travelled, so far. I figure the bend – and what we hope will be a campsite – is a kilometre away, little more. There's half-an-hour, maybe forty-five minutes, of light left. In ordinary walking conditions that would be plenty. But the terrain is constantly changing and distances on the river, particularly in the gloom, can be deceiving. We follow a cattle pad along the edge of the water. Light rain passes intermittently over us.

The river splits. We hug the shoreline of the rightmost watercourse, but with each step the bed grows muddier, and the vegetation thicker. I'm vaguely aware of the bank on our right steepening, but I'm no longer thinking clearly, am focused on reaching the bend ahead and somehow finding a way to scale the bank to our campsite. Then the tangle of ti-trees and callistemons gives way completely to a body of muddy water, and our way is blocked. I assume it's a lagoon, that the swollen river has backed up, and we'll just have to skirt the lagoon to continue on our way.

As we re-enter the forest of ti-trees and begin climbing over fallen logs strewn along the water's edge, I realise it's not a lagoon at all, but a creek flowing into the river, and we're going to have to make our way up it until we find somewhere to cross. There's no passable route along the sheer bank of the creek itself, so we backtrack and climb up into the paddock high above the river, and work our way along cattle pads keeping the creek more or less on our left, peering down from the paddock to the creek as it cuts its way through a steep gully, still too wide, still too deep. Eventually, after half-a-kilometre, the creek begins to taper. We think it's narrow enough to leap across, so we scramble down the gully wall to the water. But it's flowing quickly. Our legs are spent. Neither of us has – now that we're down here with full backpacks, waterlogged boots and heavy, wet clothes – the strength to try to jump the creek. Perhaps we can wade it. I measure its depth with a walking stick. Thigh-deep. It's almost dark now. Half-a-dozen Charolais observe us from the paddock on the other side. Bugger it. I go first, bracing as I step into the coursing stream. Ian follows. He stumbles but doesn't fall.

We make camp just north of a stand of trees, the swelling river below us. The ground is wet, but the rain holds off as we pitch our tents in the near dark. I lean my walking sticks against an old fence post for the night. There'll be no campfire this evening.

I join Ian in his one-person tent to prepare our meals on our gas cookers, dehydrated pasta for him, cauliflower and dhal for me. I lie on my side, propped up on an elbow, boots and gaiters and half my legs poking out the door of his tent. We busy ourselves with the necessary and the immediate: screwing the gas canister to the burner, clicking its moulded plastic legs into place, pouring half-a-litre of water into the pot and setting the pot and burner onto flat ground. Turning the gas knob with my left hand, flicking the lighter on with my right. And in the two minutes it'll take for the water to boil, tearing open the top of my meal pouch and arranging the folds at its bottom into a rudimentary, but effective, stand. Then pouring the boiling water into the pouch and stirring, pressing out the air, and sealing the clipseal for fifteen minutes. We talk about the Bureau of Meteorology's increasingly urgent weather warnings for the valley and discuss tomorrow's route – a shorter day to a cabin where we can dry out – and agree to a day-by-day approach.

The hot meal is rejuvenating. I return to my tent, unclip my gaiters, undo my bootlaces and strip out of my wet clothes before folding and placing them above my boots under the alcove of the tent. I change into my second set of clothes, zip up the tent, climb into my sleeping bag and lay my head on my rolled puffer jacket, which has doubled as my pillow each night. By 6.30 pm, I'm sore and tired but fed and warm and so, quickly, asleep.

Day 10

Mount Beppo to Caboonbah

Distance: 6.52 km

Evening Camp: Place of the refuge on the hill

Dawn works its way by osmosis through the low grey cloud hanging above us. Though weak, the light is even. And though the rain has been falling constantly during the night, it is beginning to ease off now. I climb out of my tent. What a spot we've stumbled upon! We're on the high bank downstream of the river's confluence with Lagoon Creek. The river is below us to the east. In the middle of the stream is an island, three hundred metres long, covered with castor-oil plants, thick, high-stalked, leaves like eight or nine-fingered hands. The channel nearest us – narrower than the one on the other side of the island – is the main channel into which Lagoon Creek flows. The secondary channel only runs when the river swells, as it did in February, as it is now.

I step across what's left of a collapsed barbed wire fence running between our campsite and where the bank falls steeply away to the water. Its strands of wire are slack and tangled and wear a necklace of flood-swept grass. So, this is how high the flood got in February. I look down at the deep brown water in the channel, thick with sediment, from the catchments of a thousand creeks and gullies. I straighten my shoulders, breathing it in. But then, in my peripheral vision, movement. Something in the river. I turn my head and glimpse, I think, a small dark triangle sinking below the surface of the water, leaving a ripple of concentric circles. So yes, something. I keep watching, and yes! Another dark shape rises slowly from the water for a couple of seconds before slipping back again. Then, a little further downstream, a third time I see a triangle of flesh rise into the air. Do I hear an intake of breath? Am I imagining it?

Lungfish. Come up to breathe as the water quality in the flooding river changes. What a flooding river taketh away, it elsewhere returneth.

We change back into our wet clothes, wet socks and water-heavy boots. We fold our wet tents. Our packs are heavier this morning. Though our clothing is cold at first, once we start walking our bodies warm quickly. As we leave the campsite a great egret flies overhead, crying. Does every greeting carry a warning too?

~

It's a beautiful misty morning. The gentle sound of the breeze rustling through the grass is our companion for an hour or so. It is quieting. Quiet that is, except for the plethora of bird calls. Ian knows them all.

We come upon a mystery. On a patch of sand on a high terrace above the river, is a large tilapia, its body picked clean, the head separated from its spine. Its mouth is agape, teeth bared, eyes missing. The scales behind its gills have begun to lift and separate. They are beautiful, thin translucent discs glowing pink or faintly green in the soft light. A dozen lie individually strewn on the sand. What has happened here? The rain has washed all bird or animal prints from the sand, and there are no traces of recent human activity nearby.

How did this fish get here? We settle on an osprey or sea eagle fishing the tilapia out of the river and carrying it up here onto the terrace to feed. The land is alive with traces of what has been, clues of what is to come.

We scramble back down the bank and step onto a mudpan, a vast jigsaw of geometric shapes formed after the mud laid down in February has dried and cracked apart under the sun. The cracks are deep, fifteen centimetres or more, with sheer walls. I peer down those little canyons to their bottom and see a layer of sand and realise that February's flood had deposited an even layer of mud upon what had been a great bed of sand. I think of the sandy beach we crossed just yesterday and am overcome by the impermanence of all things. That the ground beneath my feet is not solid. What today is sand might tomorrow be mud. Time itself cracks open beneath me, hours merging with eons. Geological time looks like this.

Look harder. What the floodwaters also carry: seeds. And I notice – because slowly I'm learning to see – that growing from a nearby crack is a little castor-oil plant, six unfurling leaves on six thin, hopeful stalks reaching out of the crevice in the dried mud. I look out across the mudpan and see now how it's dotted with green. See the weeds growing out of the cracks: castor-oil plants, nutgrass, Noogoora burr, the odd thistle.

I think back a couple of days to the forest of green panic on the riverbed upstream from Harlin, too thick to penetrate, and understand now how that came to be there.

I think ahead. If the river continues to rise in the days to come and covers this mudpan, these weeds will drown. The cracks will be filled with another deposit. Perhaps sand, perhaps silt. The water will recede. The mud will dry and crack. There will be beauty. Something will grow. A pilgrim will follow a river. If not this, another. There is no end to human wondering.

~

An unusual contraption rests on the high bank beyond the mudpan, a metal bin mounted on large timber beams, a rusting piece of machinery from a mining or sand-and-gravel extraction operation a time long gone. Someone has erected a stone cairn on top of the ridge overlooking the river with the words 'J— and P—'s special place' inscribed on a plaque. Half-a-dozen seagulls patrol a little sand island in the middle of the river. Crows caw from the far bank. Ian follows the croaks of a rocket frog, keen to spot it, but it evades his attention.

An intermediate egret stands in the shallows, its eye fixed on something in the water more important than two approaching men. It readjusts its head and long neck to follow the movement of a fish just out of reach. But its eye remains fixed. Ian and I stop to watch. Wait, wait. That such stillness can presage such kinesis. Because suddenly the bird snaps forward and spears the surface of the water with its beak. In that moment there is only yes or no, success or failure, a binary simplicity. When the egret lifts its head there is nothing in its beak. Time and place are released. The egret pulls its long neck back into itself, and takes to the air.

The river is wide, wide. Widening before our eyes as it fills.

We track the waterline. A fishing lure hangs in a catch of debris. A slice of rusting metal burrows into the sand. We step onto a basalt bluff and scramble along rough ledges; reaching, levering, shifting weight. There may or may not be a route through. The cabin must be close. A pair of black swans follow our endeavours, grace to our ungainliness. The cliff becomes steeper the other side of the bluff. We must find another way, must look for passage away from the water. We retrace our steps, relocate our handholds. A cascade of dislodged pebbles clatters down the face and into the river.

We follow a narrow animal track up through the lantana to the top of the bank and skirt the cliff-line on our left. Michelle and Charles' cabin must be very close now. I check my GPS. The satellite maps tells me the cabin is on a hill, set back from the water, out of flood's reach, a ridge, a gully and a paddock or two away. We decide to approach it cross-country.

A colony of wild melons, ripening, the most mature slashed open and emptied. Pigs or bandicoots. We come to a barbed wire fence at a sturdy corner-post. Climb up and over. On we go, thrashing our way through a paddock of spear grass. The wet is a blessing. If we'd had to cross this paddock in the dry the spears would have sliced our legs.

Ahead, Ian waits until I join him, then motions for me to bend close to a clump of broad-bladed grass he's been inspecting. I look but can't see what he sees. He carefully moves a blade with the back of his hand. The leaf is heavy with raindrops that funnel and run with the movement. Ian waits until I make it out, a small dark shape clinging to another blade of grass. A butterfly, medium-sized, dark-winged with dark circles near its margins. An Evening Brown, one of the few up here in this rural valley that is active in winter. Lovely.

The gully ahead of us is thick with marsh grass. We slosh across the boggy ground, and then stride up the hill. Four cows are resting nearby. One rises to its feet and eyes us steadily. There are other signs we're approaching a farmhouse: four-wheel drive tracks, taut fences, phone lines and the top of a row of homestead trees visible over the hill. We follow the track. The rain has held off, but the closer we get to the cabin on the hill, the more expansive the view of the river valley, and the better we see the dark clouds forming an impenetrable blanket from horizon to horizon.

~

We unpeel ourselves on the porch outside the cabin door. Walking poles. Hat. Backpack. Gaiters. Boots. Socks. We hang our backpack covers and rain jackets over plastic chairs, then wash our boots and gaiters under an outside tap, so much mud and grass and grass seeds coming away. Inside, the cabin is clean and warm and cosy. Luxury.

~

We arrange our wet gear around the heater. I elevate my sore feet and it is a relief.

Over dinner we remember the birds we've seen during the day:

- crows
- rainbow lorikeet
- pheasant coucal
- fairy martin
- red-backed fairy wren
- osprey
- white-breasted wood swallows
- seagulls
- a group of five gull-billed terns
- cormorants
- straw-necked ibis
- great egret
- intermediate egret
- pelicans
- a pair of black swans
- galahs
- sulphur-crested cockatoos
- brown quails.

Outside the world grows wetter. The drumming on the cabin's corrugated iron roof becomes louder. I turn on my mobile phone. It erupts with the pings and vibrations of concerned friends. The forecast for the valley is miserable. The flood-wary premier is issuing stay-at-home warnings. What are our plans? But Ian's phone messages are freighted with additional worry: his wife and sons have come down with the virus that's been circling the planet. His older boy has been delirious and his wife hasn't slept. Might he have the virus as well? How is he feeling? The question splits in two before he can answer it: first, does he have any symptoms? But second, heavier, how does he feel about being on the river while his family is sick at home? I give Ian the analgesic tablets from my first-aid kit for the night, just in case.

Day 11

Caboonbah

Distance of circuit: 3.61 km

Evening Camp: Place of the refuge on the hill

The sound of rain on a tin roof is the stuff of song. Of lazy dreams and autumn love and lines of poetry. Fresh lines, ancient ones. I wake, register the rain on the roof, fall away again. Wake to the rain's steady beat once more, and once more return to sleep. But the night is at work, as night always is, piecing together the day just gone, composing the one to come. A new syncopation, three pulses: the rain, my blood, my brother's breathing.

I pad softly from my bed at a late first light. Ian is still asleep, but beside his head four tablets are missing from their alfoil pop-sleeve.

Quietly as I can, I slide open the glass doors. All the sounds of rain pour in, drumming and slapping and splashing. Rain on roof and deck and soil and sand and rock. Every bared surface, every angle, every texture. Plants at every stage of their growth: leaf, bark, branch, stalk, seed. Though I may not be capable of discerning it, I am listening to the sound of rain on kangaroo fur, on feathered emu back, on turtle shell and butterfly wing. Rain on the surface of a river. Falling rain is as unique as the scape upon which it falls, and no-one in the history of creation has yet heard what I am invited to hear.

Today we'll rest.

Building in rest days on long hikes is important. It's a chance to give your body a break, and to tend to your gear – to change bootlaces, or wash clothes, or repair fraying straps. And, I think, to update your journal. Because trekking in the rain changes how you record what you experience. You can't dictate into your phone's recorder while you walk without risking damage to the device. You can't take out your notebook to jot something down that you don't want to forget without it getting wet and the paper spoiling, your notes smudging and running, indecipherable. So you have to wait until you're safely dry in your tent at night. Even then the rain changes things. Walking in the rain takes longer than in the dry, which means – if you have a destination to reach – you're out on the trail for longer each day. Then, by the time you're cocooned in your tent, it hits you how much more walking in the rain all day takes out of you – you're wet and carrying more weight, and the hood of your rain jacket is pulled close over your face so your field of vision has shrunk, and you no longer have access to your periphery, and you have to concentrate where you're placing each foot to avoid slipping. You lie down on your mat and take out your pen and journal and prop yourself up on your elbow to write, and you feel, suddenly, exhausted.

~

Sure enough Ian felt feverish through the night, meditated his way through a headache and is now feeling better. The canvas chairs on the deck are damp from the gusting north wind blowing rain in from under the roof. It's still raining heavily. The hills on the other side of the river that we'd seen yesterday are veiled by cloud. We take stock, and begin to think ahead.

Michelle and Charles' place is at Caboonbah. Half-a-kilometre or so downstream the highway – between Kilcoy to the north and the regional administrative centre of Esk in the south – crosses the river at a place called O'Shea's Crossing. From there, the river flows down to its confluence with the Stanley River, before bending right around a large peninsular, and then, after another couple of bends, flowing into the waters of Lake Wivenhoe. Wherever it may be that a dammed river ends, and the lake that its damming has created, begins.

Imagine Lake Wivenhoe as the bulge in a snake's belly where it has swallowed a giant rat. Lake Wivenhoe is young, created when the river was dammed in 1984. Lake Wivenhoe is vast. It's fifty-four kilometres from O'Shea's Crossing to the dam wall. It's two Sydney Harbours big.

Not that long ago – before the river was dammed and the valley flooded – what are now the shores of Lake Wivenhoe were hillsides, or cliff-faces, or paddocks. Now the water that presses against the land doesn't flow. It sits. Or it imperceptibly rises as the reservoir fills, or falls as water is gradually used. But there is none of the life and energy of a river. The shoreline of the lake is naked by comparison, dulled, silenced.

~

We've been altering our environment since the beginning: digging earth and moving rocks and felling trees and burning grasses. Sometimes lightly, sometimes boldly, other times brazenly. The alterations may be temporary, but often enough they're permanent – at least by human scale. And we've been redirecting the flow of water since time immemorial. For fun as much as out of necessity. There's some quality of water that invites play. As children at the beach we build walls of wet sand to hold back waves. Or dam streams with stones and sticks to better catch fish or tadpoles, or simply to observe the flow of water, to understand or wonder.

Though one person's play might be sacrilege to another.

Changing a river's course is a story as old as antiquity. I think of how Hercules was tasked with cleaning out King Augeas's stables, with their thousands of cattle, in a single day – the fifth of his labours. Hercules thought nothing of rerouting two rivers, the Alpheus and the Peneus, so their waters would flow through the stables, washing them clean, achieving the impossible by his ingenuity. But the Alpheus and the Peneus were sacred rivers, and I wonder if there's a lesson in the twist to the myth. Before cleaning out the stables, Hercules had struck a bargain with the king: he'd be entitled to ten per cent of the king's cattle if he could muck them out in a day. Because Hercules had performed his labour for a financial reward meant it didn't count, and he'd have to undertake an additional trial. His efforts were wasted. Radically changing the course of two rivers was, in the long run, futile.

~

Highways run either side of this man-made lake, formed by interrupting the flow of this river. The Brisbane Valley Highway is on the right, western shore, passing through Esk, while the Wivenhoe-Somerset Road on the eastern shore threads its way between the lake and the flanks of the D'Aguilar Range. They're both indecisive roads, uncertain about whether they're attracted to, or repelled by, the dam's water. They come and go. They skirt the lake shore at times, before unexpectedly veering away, leaving it kilometres distant, before turning back, almost kissing it until, yet again, they turn away.

Whatever this river pilgrimage might be, I must experience the lake's long shoreline, the river's drowned self. One way or another, the lake must be faced.

But now the lake is filling, and filling fast. Water is pouring into it from the river, from local creeks and straight off the saturated ground, water running in sheets directly off the flanks of hills into the lake.

Our planned route around the western side of the lake – sticking as close to the water as possible, between the shore and Mount Esk and in doing so bypassing the township of Esk – is now impossible. Two of the creeks feeding the lake, Coal Creek and Esk Creek, are flooded. There are only three alternatives to that original plan: walking along either of the two highways down the sides of the reservoir, or taking a boat the length of the lake. Both highway routes are about the same distance – a little over fifty kilometres. The eastern highway has less traffic, and might be more picturesque than the western route, but the western highway would allow a stop and resupply in Esk if necessary.

We gather information. The Bureau of Meteorology has issued a minor flood warning for the upper Brisbane River, and sure enough when I return Ali's call she tells me the river is up and she and Duckadang are now cut off. Gregor's Creek, which Dominic and I crossed four days ago, is rising quickly. Graham, when I ring him, is curious about what we have in mind. Do I realise that the highway into Esk from the north has been cut at Gallanani Creek? Ian confirms from the government's official road closure webpage that the highway is closed. So walking down the western side of the lake is no longer an option, at least for now.

Might it be feasible to travel by boat? I speak with Cheryl, a shire councillor who has been a tower of support in planning the walk, and who has a couple of suggestions. I leave a message with a boat-hire operator who rents out tinnies from Somerset Dam, on the Stanley, before getting on to Lyle, a member of the Somerset and Wivenhoe Fish Stocking Association. He listens carefully as I describe our dilemma and then ask him whether he thinks boating down the lake is possible. Do I mean besides the fact that the dam's operators have issued a notice saying no boats are currently allowed on the dam? Yes, I mumble, leaving that little obstacle aside.

I am neither a boatie, nor a fisherman. Lyle gently explains the facts of life. I appreciate Lyle's kindness. There is no public access for boats at O'Shea's Crossing, or anywhere near Caboonbah. The nearest is Hamon Cove, thirty kilometres down the lake, so that's the closest official place we could get in. But even if we could somehow find an unsanctioned access point nearby, it's a long journey, and we'd need a big boat, not a tinnie. We'd need a fair bit of fuel, and it would take a while, because the speed limit on the lake is six knots. Lyle's got a boat, a two-person run-about, not big enough for a whole-of-lake journey. And, Lyle repeats, completely fairly, am I aware that no boats are allowed on the water in a flood?

By now Lyle is invested in solving our dilemma, and wants to help. The creeks are up, so walking Wivenhoe's shoreline is out. I say that Gallanani has cut the highway down the western side of the lake. Of course Lyle knows that. He recommends, instead, the highway down the eastern side. The only creek that might breach that highway is Northbrook Creek which comes down off the D'Aguilar Range. Sometimes it floods, but it's currently open, and the road on that side is pretty much an all-weather road. Ian confirms from the official road closure website that the eastern highway is open.

So then. It's settled. The eastern highway route it is. We'll leave tomorrow.

~

The teeming rain eases to a spit midafternoon. We take the opportunity to stretch our legs – a short loop down to the river and then along the bank to O'Shea's Crossing, before returning to the cabin by the highway to the front gate and up the long driveway. Though the rain is light, water is still pouring off the hills. The cattle pad we follow to the river has been transformed into a narrow muddy creek. A tribe of magpies follows us some of the way down, stopping at a *Eucalyptus tessellaris* but coming no further. We pause too, when we reach the edge of the high bank. I barely recognise the river before us. It's higher, faster, darker than it was only twenty-four hours ago. And now it's carrying debris downstream, small objects – twigs, branches, islets of grass, plastic bottles and other refuse – but also larger items, logs and plastic tubs. A pea-green upturned feed bin swirls gaily past, missing only, I can't help but think, an owl and a pussycat.

A skin of debris.

Where has all this come from? Surely February had already flushed a decade's worth of debris out of the waterways. Is this then just the last three months' worth?

We stride out along the bank. Even so, the debris outpaces us by a factor of three. However full Lake Wivenhoe is downstream of here, it's still filling, water is still pouring in.

We count weeds as we push our way down the bank to the bridge. A fringe of dead lantana, drowned in February. Potato vine, with its purple flowers, draped from branch to branch, decorating the dead lantana. Swathes of felt-haired wild tobacco, and later, a wash of castor-oil plants. A three-centimetre-long Noogoora burr catches on the cuff of my left gaiter, and rather than toss it, I slip it into the pocket of my rain jacket. We start seeing the full array of weeds from my childhood, the memory of them sweet with nostalgia: narrow-leaved cotton bush with its milky sap and air-filled seed sacks that you'd dramatically pop beneath your heel, clutches of stinking rogers, the dreaded cobblers' pegs with their short black needles that stick rather than prick. We climb up out of the riverbed at the bridge, but don't cross. Our little weed-spotting game had been fun while we were playing, but now that we're standing on the bridge surveying the river and its bank and all the country sweeping off it to the south, something sours.

'From where we're standing, how many of the plant species that we can see are native?' It's barely a question, more a despondent rhetorical sigh.

But Ian takes it seriously. He looks out at the country, slowly turning through the points of the compass. He sees more than me. I do what I can to answer my own question, but it's the feral plants that crowd out the landscape, suffocating my sight. There are all the introduced weeds that have invaded the waterway. More on the verge of the highway, a contest between Chinese elms and Leucaena, introduced as a fodder plant, now invasive. The grass in the paddocks here are mostly all alien varieties, mainly flowering Natal grass, but some spear. The eucalyptus trees appear to me, in this moment, as remnant giants, islands of ancient knowledge surrounded by a rapacious, unstoppable tidal present.

'About ten per cent,' Ian reckons.

~

By eight o'clock Ian is lying on his back on his sleeping mat. His hands are folded over his chest, his eyes closed. He coughs intermittently. He has a temperature and pain behind his eyes. I bring him water, and more analgesic. The virus is upon him. He needs to sleep.

No matter how he might wake up in the morning, it would be crazy to set out on a twenty-five-kilometre walk in the wind and wet halfway along the eastern highway. I text Michelle, who says we can stay in the cabin for as long as we need. Tomorrow will be another rest day.

Day 12

Caboonbah

Distance of circuit: 4.74 km

Evening Camp: Place of the refuge on the hill

Ian feels better this morning. Even so, we decide to put another twenty-four hours between the virus and a big day's walking. The western highway is still cut at Gallanani Creek, and the eastern highway is still open. Rest, little brother, rest. Though I do too, my bruised heels propped on a pillow on a chair. I write. We read. Every now and then we take turns to get up and scan the sky above the river valley. The cabin is a godsend. But the cabin is also small. By midmorning Ian is keen to test his health.

~

Through the mist a whipbird calls. It has no mate to answer it.

We return to O'Shea's Crossing and this time walk out onto the bridge to take in the swollen river. Debris continues to flow downstream beneath us: gatherings of grass, islands of foam, limbs of trees. I watch the eddies in the current. Each swirls clockwise.

In the mist around us, hundreds of martins and swallows swerve and dive and twist and dip. After watching them for a while, what had at first appeared to be a chaotic jumble of small darting bodies begins to take form. I begin to differentiate the martins from the swallows, begin to see the different sizes of the birds, begin to see their slightly different arcs. And begin to see that the martins are, predominantly, on the downstream side of the bridge, the welcome swallows on the upstream side. Have they reached some territorial accommodation with each other? Or does the bridge disturb the natural flow of the air, creating different air currents downstream of the bridge, and is their darting and turning and dipping against the grey sky and brown river, a preference based on a fine difference in the respective abilities of the two species to manoeuvre in currents I can't see?

I stand at the very centre of the bridge, the brown water coursing and eddying below, a universe of birds swirling above and around, the pulsing mist enveloping me. I close my eyes. I listen.

I am on a balcony, looking down at a mosaic of the river at the centre of the world. 'My soul has grown deep like the rivers,' I hear Langston Hughes singing as he calls up river after ancient river, the Euphrates and the Congo, the Nile and the Mississippi. When I first read that poem, 'The Negro Speaks of Rivers', I felt Hughes was speaking for me too, or at least that part of my being that responds to the ancient song of river. Call it soul.

Years ago, in Harlem, at the Schomburg Center for Research in Black Culture, I took in the grand mosaic of tiles laid into the floor – a tribute to Hughes and his poem. I was pulled back and forth between all that is common to every river, and that which is unique to each. As the Schomburg Center describes it, the cosmographic mosaic maps Hughes' rivers, 'weaving a web of connections between people of diverse cultures and backgrounds, the past and present'. But as I study the mosaic I realise other rivers have been added to the ancient ones Hughes sings about. And I see, there in Harlem, the Murrumbidgee of my own continent. One of *my* brown rivers. It was enough to make a soul long for home.

~

There's a constant flow of information about the river, and the flooding. We check and recheck the Bureau of Meteorology's website: its flood warnings for the catchment, rising like piano scales, minor, moderate, major. Record rains are now forecast for May. The BOM's radar images show bands of heavy rain sweeping in from the northeast. I flick across to the government's road closures website. Red pins on the map mark each closed route. The number of pins is increasing by the hour, a spreading contagion of them. The shire council has CCTV cameras mounted at half-a-dozen sites prone to flooding. We can see live online what each of those cameras sees.

We watch the floodwaters swirling over the Brisbane Valley Highway at Gallanani Creek. Upriver, crossing number 1 at Mount Stanley is flooded. Downriver bridges are falling like dominoes. The premier continues her public warnings for heavy rain in South East Queensland. Evacuation centres open in half-a-dozen locations in the river valley.

I ring a property owner downriver of the dam wall. She's looking out her window at the river as we speak and guesses the dam operators have done a water release because the river is three times as wide as usual. They'll do that, she says, the operators – open the dam gates if the water level gets dangerously high.

Messages keep coming in from concerned friends. We're safe, we tell them. We're holed up in a cabin near the top of Wivenhoe. We're fine, really.

We scroll through the websites one last time, checking the latest information before we turn in for the night. Gallanani remains flooded, so the western highway remains impassable. There's also an update on Northbrook Creek that we don't want to hear but which doesn't surprise us – it too has now flooded. Which means the eastern highway is also out.

The highways on both sides of the lake are cut.

We're not going anywhere tomorrow.

Day 13

Caboonbah

Distance of circuit: 12.62 km

Evening Camp: Place of the refuge on the hill

I pull the brim of my hat down, zip my rain jacket to the neck and thrust my hands deep into the side pockets of my jacket. For a minute or two, at least, I've subdued the wind. I can feel the Noogoora burr I'd collected yesterday, still lodged deep in the right pocket of my jacket. I roll it between thumb and forefinger, its hard oval casing covered in sharp spikes. A little warming finger massage to keep the weather at bay.

Ian joins me at the cabin door. Downriver of here the Brisbane runs straight till the Stanley joins it from the left – the north – after which the river bends right, straightens, then bends right again, forming a long peninsula of land. If we end up walking either highway, we'll miss the peninsula. I've walked it before on a planning day, have stood looking across at the mouth of the Stanley, been thrilled. We can't miss the peninsula, and the Stanley's confluence with the river, and this beautifully proportioned sweep of river. It's a twelve kilometre round trip from the cabin. No matter how heavy the rain, we can't miss the peninsula.

We step out, walking quickly cross-country through the home paddock towards the highway, over the boundary fence and then out onto the wet bitumen strip. Because the highway is cut both to our left and right there's no need to look for traffic. We've got this stretch of highway entirely to ourselves. We cross, and then climb through the boundary fence on the other side of the road.

The earth is completely saturated. It can absorb no more. All the falling rain runs in sheets across the ground, feeding creeks and gullies, forming new ones. The land is alive with rivulets and bogs and shimmering plains of water. We reach the swelling river and set off along the bank. But every fold in the earth has become a rift and what would usually be creeks are now water-filled gorges, impossible to pass. Cascades pour off the high bank. We climb back up to the top and make our way along the heights, the wide brown river on our left, folds of waterlogged grazing country to the right.

~

When I walked this peninsula last winter, a reconnaissance of sorts, the bed of the river was wide and the walking easy. Melons grew bountifully in the sand. Swans and pelicans floated gracefully by. The blue sky was just as generous, accommodating cloud and bird and breeze. When a sea eagle lifted from its high nest in a eucalypt on the far bank to find the first thermal of the day I fell into a reverie. I'd left the highway and its traffic behind. The powerlines had marched away out of view. There were no houses in sight, no farm sheds, no fences, no water pumps. I sat on a log, immersed in the natural world. There was not a trace of another human within sight, not a shred of evidence of their existence. Or so I imagined.

I looked up from my dreaming and saw the moon at first quarter looking down on me.

'No,' the moon seemed to scold, 'you misunderstand. There is nowhere to go. You are human too. You and your brothers and sisters are fully and completely a part of this. Call it complicity or call it belonging, whatever you wish, but I see your intimacy even if you do not.'

It was the strangest feeling. That it was the moon – ethereal, otherworldly – that had snapped me from my whimsy. The moon from its distance, taking in me, and the river, and the farms and powerlines over the hill behind me, the entire river valley, the earth and my eight billion human relatives, all my fellow creatures.

~

Ian and I talk about our childhoods, five years apart, from a distance of decades. Both the shared memories, and the experiences of one which are unrecognisable to the other. We know the course of each other's lives well, but as the years have gathered, some milestones have fallen by the wayside. We tell stories and piece our lives back together. We talk about the creeks and gullies of our youth. Of exploring the bush behind our home. Of the pioneering mythologies – and virtues – sewn into the fabric of our childhood: adventure, courage, resourcefulness, resilience, determination. We talk about the creeks we've taken our own children to, and how their explorations are animated by slightly different values. How my boys' backyard creeks – running through Melbourne and Brisbane – offer the same comforts as our childhood one's did.

~

The banks of the Stanley where it joins the Brisbane on the other side are indiscernible through the mist. What I can make out is entirely changed from what I'd seen from this same spot last year. Then, the mouth of the tributary was distinct. Now there's no mouth to see. It's a lake out there, a sea.

I wipe my glasses and dry them as best I can. A mob of eastern grey kangaroos bounds away through the wet grass.

We walk to the end of the peninsula. The sweep of the current bites into the bank at one point, forming a secluded cove. But the beach of this strange inlet is composed of debris, and at its high-water mark a pontoon lies stranded. It must have broken its moorings somewhere upriver and washed down in February where it's become marooned, lame now, useless, but in perfect condition. Most of the pontoons that broke free from their riverfront properties in the city during February's flood are of a different design – carpeted aluminium decks floated by white polystyrene foam slabs. But this is a custom-built bush pontoon: six sealed blue plastic forty-four-gallon drums beneath a bespoke-welded aluminium frame, with a deck of ten lengths of timber bolted to the aluminium. It's a raft, waiting to be refloated. I climb up and survey the debris field from its deck. The flood's currents have delivered a beach of logs and branches and fence posts and plastic refuse thirty metres wide. It's a scene of devastation, dead, like a lava field. But, up here on the deck of this raft, I float above it all. I dance, my face to the lashing rain.

We must, eventually, go. I climb down. We set out again, sinking with each footstep, walking in slow motion, this alien landscape. We cross the lifeless plain, climb above the flood mark, and make our way up the side of a bowl. We look for raised ground to place our feet – mounds, tussocks of grass – but they are mirages. The entire hill is a cascade, splashing our thighs, spraying up into our faces.

At the top of the ridge we look south to the swollen river and its variable inlets and all its back-filled lagoons and the great lake its damming has created. So much swelling water ravenously gobbling bank and shore. There are fingers of land down there – thin, low, distant, eucalypt-covered peninsulas – but the river is destined to break through them, sweep them away, today or in a thousand years. On a day such as this it feels like it just might rain forever.

We turn away, turn for home, picking a fence-line to lead us down the centre of the peninsula, back to the highway. My glasses are wet again, but now they're also smeared from trying to dry them with the corner of my shirt, itself drenched. I spot a mob of emu running down the side of a mist-shrouded hill to our right. After the eastern greys earlier this morning, I want to say it aloud: this is still kangaroo and emu country. But I wipe my glasses and look again, and they are deer, long-necked, ethereal, floating through the grass. We lose them, and walk on around the hill where we pick them up again, through a gap in a stand of eucalypt, still running. They are their own mystery.

The fence leads us up to a set of old split-railing stockyards. Ironbark. We climb over the rails. The ground inside the yards is a mire, so we skirt its edges, our fingers guiding us along the lengths of the hand-sawn timber. Olive and pink lichens pattern the rain-dark railings. A bramble wraps itself around a corner post. A giant ants' nest grows over the low rail. One way or another these yards will last forever.

~

The black dye from my Akubra hat is beginning to run and my left ankle is growing sore once again. I limp the last couple of kilometres back to the cabin, despondent. These days of unexpected rest haven't fixed my ankle. But then it comes to me, a revelation. I realise that after two weeks of walking mainly on the right bank of the river, I've been walking on a bank angled the same way, footstep after footstep. My poor left ankle has been supinating nearly continuously for thirteen days and the ligaments and muscles are simply over-stretched, tired, done.

~

Back at the cabin, soaked and muddy, we reprise the routine of undressing and cleaning we'd used when we first arrived here three days ago. Washing down our boots and gaiters and bringing them in to dry, hanging wet clothes over a makeshift clothes line. I recover the Noogoora burr, my little meditation bead, from the pocket of my jacket and lay it on the coffee table beside my notebook.

'I can't remember,' Ian mulls when he sees it, 'if the Noogoora burr arrived in an Afghan cameleer's saddle, or with cotton seed.'

He picks up his phone to research online.

'Mississippi cotton,' he reads from an academic article, and then, his voice lifting, shot with surprise, 'but get this. It's called "Noogoora burr" because when it was first located in Australia, it was found on a property called Noogoora that had been used to grow cotton. Guess where that property was?'

'Nooooooo ...'

'Yep. On the Brisbane River.'

We spend the next hour online trying to learn something more about this property, Noogoora, but find nothing. It's as if the property has vanished, to live on only in the local name of an introduced burr.

~

Afternoon falls into evening. We eat. We check the flood levels, again and again, flicking between Gallanani Creek on the western highway, and Northbrook Creek on the eastern route. We decide that whichever of the two creeks recedes first, will determine which route we take. We can observe Gallanani ourselves by camera, whereas we're relying on updates about Northbrook. We tell ourselves Gallanani is dropping, but don't trust what we think we're seeing. We distract ourselves by reading or writing or sleeping, before checking the camera again. It becomes a game we play with ourselves. But by nightfall it's certain: vehicles have begun to get through again.

It's eighteen kilometres from the cabin to the Gallanani Creek crossing. We decide that if the highway is still open when we wake tomorrow, we'll go for it.

Day 14

Caboonbah to Hamon Cove

Distance: 32.83 km

Evening Camp: Place of the rising lake

The day breaks overcast, but it is not raining. Gallanani, eighteen kilometres down the highway, is still open, and if the rain holds off surely it will remain open, at least for the morning, long enough for us to get through. We step out into the grey day, close the cabin door behind us, walk down the long gravel driveway and step out onto the Esk–Kilcoy Road. We are leaving the river in order to return to it.

A light breeze blows from the northeast, over our right shoulders. Our packs are not yet heavy. The habits I'd developed while walking on the riverbank are no use to us now. Walking highways demands different routines. We walk on the right-hand verge of the road, walking towards the oncoming traffic. But the verge is narrow, and we have no real choice but to walk on the road in the traffic lane itself. Though we shouldn't have difficulties seeing vehicles coming towards us, we agree to call out each oncoming car to give us time to step off the road. Ian's backpack cover is a light blue – more visible to motorists in the grey conditions than my black one – so he takes the lead. A crow arks at us from the river side of the highway. I take it as a positive omen.

We pass the old Caboonbah Homestead on our left, the family home of Henry and Katharine Somerset, built in 1889 and destroyed by fire in 2009. Its gates are locked to the street now, authorised visitors only. 'Caboonbah' is said to be derived from the Kabi Kabi words 'cabon gibbah', big rock, the house built on a high rocky bluff, the site chosen because it was safe from flood. On the other side of the high chain-mail fence, classic colonial homestead plants grow in the garden, cycads and jacaranda and figs. They, at least, survived the fire.

~

It's here, at Caboonbah, that a local legend was born.

The story goes that a horseman, Billy Mateer – a 'good game stockman' – outrode a cyclone in 1893 to warn the good folk of Brisbane town about a wall of floodwater making its way towards the city. The story is a mix of Paul Revere's midnight ride, the legend of the first marathon runner who ran to warn the Spartans about the approaching Persians, and the tale of the little Dutch boy who kept his finger in the dyke long enough to save his townsfolk from flood. It's a good legend, and deserves embellishment in the years ahead.

The local nobleman, Henry Somerset, spied the floodwaters from the verandah of his homestead perched high above the river. If the downstream capital was not warned, it would be destroyed. But the telegraph line this side of the river was down. The river would need to be crossed, and from there the message carried over the range to the next telegraph office. So nobleman and horseman ran down to the stables, unhitched the best two horses and lashed them to a rowboat on the banks of the transformed river, in order to swim the fine creatures across. Only one of the horses made it, Lunatic. So Billy saddled him up. And ride he did, gorges as deep and black as those navigated by Banjo Paterson's 'The Man from Snowy River', any slip death. Only that hero was riding in perfect alpine weather. This was cyclonic. Up the D'Aguilar Range across flooded creeks, Billy rode. Through the thrash of storm, the fate of the city on his shoulders. Down the other side again, over fallen logs, down rough and broken ground, wet and deathly slippery. Imagine him spent, relieved, arriving at the telegraph office at North Pine where, muddy and sodden and panting, he handed Somerset's note to the station-master to telegraph the flood warning to Brisbane where on 19 February 1893 the floodwaters at the Port Office gauge reached 8.09 metres.

It's a ride that may yet become myth. It has all the elements. The gods of nature. Their wrath and their power. The destructive power of a river in flood. A race against time. A hero. A city saved.

However, Somerset had sent another telegram two weeks earlier, when the river flooded for the first time that February. It had reached the Under Secretary for the Post Office in Brisbane: 'Prepare at once for flood. River here within 10 ft of 1890 flood, and rising fast; still raining.' The message got through but was ignored. This was Somerset's second attempt that February to warn Brisbane. And this time? By one version of events, Lunatic got bogged and Mateer had to walk the last leg into North Pine, Somerset's warning arriving in Brisbane about the same time as the river, too late to save the city. By another version – Mateer's – he delivered the warning and it was only on his return journey his horse got bogged. Evidence that the telegraph got through and cityfolk heeded it to save life and property? Scant.

But that's never the point with myth. Myth answers need. If the need is strong enough, there'll be poems and paintings and stories and songs about Billy Mateer.

~

From the river to the road. From water to bitumen. From shoreline to painted white rumble strip. We've left the domain of flesh and entered a zone built for machines. But this morning, though the road is open, we're untroubled by cars. We pass boundary fences and driveways into properties. Guardrails and speed signs. A blue New York Yankees cap has blown off a dashboard onto the verge. A plastic doll sits straight-backed on a rock, propped up by a single brick, wide blue eyes and long lashes. We pause before a simple cross screwed to a tree, memorialising John, 25 Feb 20, in yellow and blue.

Fingers of the lake touch the highway in places. Nameless creeks demand attention, roaring with water as we pass by. The road bridges Coal Creek, mud-brown today, but its local history is carried in its name. We step off the road for a ute, the first vehicle of the morning, a good sign that vehicles are getting through and that the highway over Gallanani is still open.

Gallanani. The word has grown huge. How often we've said it these past four days. It's a creek crossed by the Brisbane Valley Highway, yes, but it's metamorphosed beyond those physical contours. Taken on its own mythic proportions. Gallanani has come to represent all that is just out of reach. It is the obstacle that can't be wrestled with. The mercurial direction, reversible at whim. It is flood and it is fate and it is hope, tired and bedraggled. Now, finally, just ahead, over the next rise, is the thing itself.

Gallanani is a creek that in quieter times flows – if it flows at all – through culverts beneath the highway. But in flood Gallanani becomes a lake, submerging banks and paddocks and highway.

Receding Gallanani is a field of human activity. Men and women in yellow vests on two-way radios. Traffic controllers and road workers and emergency services personnel. There is machinery being loaded onto the backs of trucks, their job done. Temporary traffic lights regulate flow along the highway. Traffic backs up, then is let through, backs up, is released once more. The traffic controller looks us up and down, bemused. She shrugs and signals us through. 'But don't dawdle.' We hurry down the highway, quickening pace where the road passes over the muddy withdrawing creek. The verge is narrower there, and we're keen to get across before the banked-up vehicles are released and overtake us. I bend to pocket a chunk of bitumen, loosed from the road by the floodwater.

~

It's another two-and-a-half kilometres into the township of Esk. The old railway runs parallel to the highway, shoulder to shoulder at first before separating, a respectful gap opening up between them, some long-ago pact between rail and road. We take the rail trail. The steel and the sleepers are long removed. Gravel crunches loud under our boots. A monolith, Glen Rock, rises out of the plain ahead, forcing the river to curve around it, away from us and from the township, bending the river to its will, separating us from our journey's companion even further.

Roll and crunch of gravel, louder than blood, louder than speech. We crunch our way towards town.

A monarch butterfly, that great wanderer, alights upon a cotton bush, its beloved milkweed. Burnt-orange wings glow in the mizzle, felted wing on flushed cheek. Familiar, but it too, is a stranger, a migrant to this land. And it too is home. Its precise passage from North America may never be known – invasion pathways sometimes leave no trace – but they probably arrived from the South Sea islands, where New Caledonia and the New Hebrides had populations before they were first spotted on the east coast in 1870, and by 1873 were 'everywhere – a sudden American invasion of the whole continent'. Perhaps ova were attached to tropical milkweed when it was introduced by the Acclimatisation Society, or individual butterflies might have made their flights across the sea on summer trade winds. A bolder hypothesis: that clouds of monarchs were swept up from New Caledonia by one of the three cyclones to cross the Queensland coast in the summer of 1870 and deposited en masse.

What weather system is crossing the coast now, pushing Ian and me off course?

Three chestnut mares stand on the wall of a small earthen dam. One raises its head, glares at us with a suspicious eye, bares its teeth, shakes its mane, gum froth visible even at this distance. But we are tiring and gravel-deaf. At the water's edge: wood ducks, Pacific black ducks, whistling ducks and teal. They break as we pass, orderly, the teal taking to the air first, the Pacific black ducks last. Duck-wing slap on water in slow motion, duck beat in air. But our walking has become unthinking and the detonation of gravel beneath our marching boots overwhelms bird and pond and sky. On the edge of town, from a backyard, I hear a familiar song. It is the first butcher bird of the journey, shoulder high in a eucalypt.

It is not quite midday when we arrive at the café with the garden in the main street. We take stock: eighteen kilometres of bitumen, one aching left ankle (now raised on a spare chair), the river-cum-lake on the other side of the monolith, out of sight, and a band of lorikeets swooping onto our table in expectation of the meal that's yet to arrive. I ring a friend, Donna, who lives in town and who'd dropped by the pharmacy before it closed this morning to pick up a brace for my ankle. When she joins us, it's with a second gift as well: an offer to lighten our load and swap our backpacks for lighter daypacks for the remaining thirteen kilometres to our campsite tonight. Oh, Donna! Let me give you a hug!

~

This second half of the day's walk, though shorter and with much lighter weights on our backs, is tougher. We're faced with traffic now that news of the highway's opening has spread, a continuous rumble of oncoming vehicles. It feels like we're walking into the teeth of a gale of cars and trucks.

I am beginning to resent the dam. It is forcing this highway upon us, noisy, hard, bare, brutal. My ankle aches. Each step is hard. Boot slam, reverberation. We take turns to walk backwards for stints, buddying each other for oncoming traffic. Ian walks backwards to relieve the muscles that have been doing all the work – the same stride-length, meter, pulse. He gives other muscles a chance to work, not get lazy. I follow his lead, my immediate aim simply temporary relief for my ankle. Light rain sweeps over us. I take a hundred backwards steps and watch where we have come from receding from view.

~

At 4.30 pm we reach the turnoff to the cove, cross the highway, and follow the access road to the lake shore. Floodwaters have spilled across the road, cutting off the picnic area. There's a four-wheel drive pulled up where the road is being swallowed, its rear door up, the lake water lapping at its wheels. Inside, the vehicle is meticulously arranged with the gear of a serious angler: rods and buckets and nets, waterproofs, bait, icebox, custom-fitted tackle cabinets. The fisherman himself is leaning against his car, lake-side, a line in. We didn't mean to startle him, but he hadn't expected anyone to come down the road on foot. He checks us out, two hikers kitted out for their own outdoors experience. He relaxes. There's a kinship among loners. He's never seen the dam this flooded, he tells us in broken English, and doesn't expect to catch anything because of the mud and all the rubbish. He's Chinese-Australian, has come up from Brisbane, can't believe how few people he ever sees up here. He fills us in on what we need to know, without us needing to ask. The gates to the picnic area close at 5.30 pm and open at 6.00 am. The ranger lives over at Logan's Inlet further along the lake shore, and won't visit again after he closes the gate. The property owner at the top of the hill is friendly.

Donna and her husband, Simon, arrive with our gear. We return their daypacks, and shoulder our backpacks once again. While the road to the picnic ground is submerged, we follow its cut along the side of the hill, and drop down onto the bitumen again where it rises out of the swollen lake near the toilet block which remains, still, above the water level.

Out in the lake stands a picnic table beneath a peaked corrugated iron roof. The tables' fixed bench seats are completely submerged, invisible beneath the muddy water. Only the table top, and the tin roof and the four posts supporting it, are visible. All else is awash. The table top appears as a raft on the choppy waters of a vast lake. But for how long can it stay afloat?

~

We set up camp among a forest of *tessellaris* saplings. The rain returns. I am asleep by 7.45 pm, having walked thirty-two kilometres of bitumen. I thought I could sleep forever, but wake at 9.30 pm, my legs twitching and cramping. It's as if being still is too much for them. I drink from my water bladder, and check the internet. The water authority that manages the dam updates its water levels regularly. At 10.15 pm Wivenhoe is at 134.4 per cent, a figure that prompts the question: How can a dam be more than 100 per cent full?

As far as complex pieces of engineering infrastructure go, dams are simple things. They store water. Most dams do it for two purposes: first, to secure a supply of water for our future use; and second, to control the flow of water that might otherwise flood whatever is downstream of the dam. Some dams have a third purpose – generating hydroelectric power – but that's another story. When the dam operator says the dam is full, it means the dam has reached a pre-determined volume of 'supply' – the water it aims to secure for homes and agriculture and industry. The rest of the dam's capacity – the difference between that arbitrary level of supply and actually being full to its brim – is for flood mitigation, managing the risk of disaster.

So the dam operator serves two gods. The god of thirst and the god of disaster. The more water the dam stores, the more the god of thirst is kept at bay. The emptier the dam, the more water it can hold back during heavy rains and the more the god of disaster will continue its slumber.

If the city has more water than it needs to supply the city downriver, that excess can be released. *Should* be released if the volume of water the dam is trying to hold back threatens to damage the infrastructure of the dam itself. Because if a dam wall fails, catastrophe follows.

To release or not to release? To hold or fold? And if to release, then how much, at what rate and when? It's the fate of a dam operator to work odds, like a gambler. But they're gamblers armed with a manual, at least the operator of this dam is. A manual that tells them when to pull the plug. Literally. A manual that, after January 2011 and the damage visited upon the city following that flood, has been pored over in courts and enquiries. A manual described by one judge like this: 'Essentially, the procedure described in the Manual involved the engineers predicting as best they could what the likely maximum amount of water was going to be' and when it would arrive in the city downriver. 'Then, appreciating that inflows might exceed that prediction, they were to regulate outflows by reference to the primary consideration as identified by the Manual.' That is to say, the manual answered the foremost question a flood engineer must ask during a flood event: 'Given the likely maximum amount of water in this dam, should I be focusing on protecting the dam structure, or should I focus on protecting urban infrastructure, or may I merely focus on protecting downstream bridges?'

But this dam operator seems to shuffle numbers, flip equations, deal fresh hands. Pick a number, a different number. Tonight its website tells me that full supply isn't 100 per cent, but rather that 90 per cent is full supply, and that above that, there will be gated releases of the excess supply.

So Wivenhoe is 134.4 per cent full, and the dam operators are releasing some of what they don't need until they bring the water level down to 90 per cent. Not too much, not too little.

But still it rains. Still water pours off the flaps of our tents and across the ground and between the saplings and the leaves of grass and into the rising lake.

I am exhausted but cannot sleep. My body is tight. I am caked in perspiration and precipitation, a potent mixture of liquids. The steady rain on the roof of my tent does not soothe.

~

Ponder too much upon the nature of river and you disappear. River becomes an impossible conundrum. A mountain exists. It is buffeted by wind and rain and storm and lifts with the movement of continental plates and erodes, but despite all it stands. A river is a dream. Today it is snake, tomorrow eel. It cannot contain even itself. It spills beyond itself, becomes lake, becomes sea. It contracts, shrinks and disappears completely in the dry seasons when sun rules. You come to realise there is nothing unusual about a dry river, a ribbon of sand. It has always been like this. The river is a parchment. Decipher its runes if you can: this charred log, that rusting railway pin, that bleached turtle shell. But the language of even those hieroglyphs is young.

There is a cave here somewhere that survived the dam, or so the old folk say. When it's given a name at all, the name is whispered – the Platypus Cave. Survived, but severed from us now. All the old tracks that once led to that ancient shelter are drowned. The cave is up there, isolated forever in its inaccessible cliff-face. One could, I guess, approach by boat. Could pull in close to the sheer rock and fasten one's vessel somehow. But I feel certain there is no route up that wall, that it is clean and forbidding, impermeable. I am equally certain, tonight, the cave holds knowledge that, though once within our ken, is now forever doomed to lie beyond our reach. That having drowned the country, we have separated ourselves forever from that which might rescue us from all the ills of our age. Memories of the cave disturb my dreams. On the walls of that cavern there's a gallery where platypus dive and dance their ochred paths across rivers of rock, a story threaded from Gondwana to Mount Glorious.

Day 15

Hamon Cove to Lockyer Creek

Distance: 19.14 km

Evening Camp: Place of the roaring waters

I wake to kookaburra. A nocturnal creature is late returning to its nest, small body on leaf litter on sodden ground. The earth moves under the weight of overnight rain. Mist drifts off the lake, between the eucalypts, curls past my open tent flap, presses against my neck. Slow and steady grows the lake. Under the cover of dark, it has been on the move, a relentless new vision of itself. Slow, but insatiable. I understand now what it was that so disturbed my sleep. The lake pulls and I succumb. I slip on my camp shoes. How many steps to the lake shore? Fewer than yesterday evening. It is rising. It comes to us. How high might a lake rise? How much water might swamp how much land? Only the lake knows, but it is inscrutable. Its task is to rise, ours to wonder and worry. Even the idea of 'shore' becomes somehow quaint, a theory with no purpose left to serve.

The lake's surface has settled from yesterday afternoon, is calmer. But the lake would fool you that it is still. A thick fog hangs above it, co-conspirator in this ruse. That kookaburra, again, now from behind. What? On the lake shore, lapping, pressing, there is a slick of grass and leaf and twig and insect wing and beetle carapace and grass seed. The lake laps at my feet. Further out, through the mist, a hundred eucalypts stand in the water. Not that long ago they were a forest. Now the lake has separated them. Each is completely isolated, alone but for their reflections, desolate, indistinct. Beyond the drowning trees, I hear a splash. Fish or god? My brother's voice from over my shoulder, soft and clear, moulding himself to the day.

~

We're out of drinking water. The water in the lake is muddy and stirred thick with sediment. We can filter it through a clean t-shirt and add purification tablets, but we wonder if the amenities block we'd skirted at the picnic ground yesterday evening might have tank water.

We climb into our wet clothes and our water-heavy leather boots, clip on our gaiters. We collapse our tents and roll them into their bags. Pack away our waterproof pouches, clip up our backpacks, slide over their rain-covers, mount our packs on our backs. As we leave the campsite behind, magpies drop down to inspect the twin rectangles of depressed grass where we'd lain last night; they are looking, no doubt, for worms. Drops of water slide down long gum leaves as we walk out of the forest, each drop detonating on our backpacks. We straddle a fence. Through the fog a wallaby lifts its head.

Out in the lake, yesterday's picnic table has now disappeared completely beneath the rising water. It's a rough calculation, but I figure that means the lake has lifted by about twenty-five centimetres. The whole lake. In just twelve hours. We stand silently, our thumbs tucked into the straps of our packs, facing the water. A large log bumps against a partly submerged fence post. It looks lost. A dead catfish floats towards the shore, its upturned eye enormous, bulging. I turn away.

The amenities block, at least, remains above the water line. I find a tap. The water is clear. We examine the reticulation system. There's a tank inside the block, being fed rainwater from the roof. Okay then. We fill our bladders. Just to be safe, we drop purification tablets in.

I check the dam operator's website. Wivenhoe is at 137.3 per cent capacity.

~

There are twenty kilometres of highway between the campsite and the dam wall. It is a trial, a penance even. I cannot complain, because I am a beneficiary of this damming. At my downriver home in Brisbane I have drunk this stored water. I've washed with it and used it to hose down my fractious kids in the heat of summer. If I had thought about it as I was filling a basin to shave I would have been ready enough to trade the drowning of fifty kilometres of river for the certainty of water during drought, assuming the terms of the trade were that simple.

So I am complicit in the disappearance of the river. If I have to endure fifty kilometres of highway ugliness, I do it in solidarity with the brutalised river.

~

We resume yesterday's practice of walking on the right-hand side of the road facing the oncoming traffic. We walk on the edge of the bitumen, or on the gravel verge if it's wide enough. We step away as each vehicle approaches. Sometimes onto the gravel verge, at other times – where there is no verge at all or it's too narrow – we have to retreat into the grass hemming the road. In places the verge is wide enough to continue walking until a car passes, before resuming our journey on the flatter surface. We treat semitrailers differently to cars. If it's a semi bearing down on us, we step well off the road, turn our shoulders and brace as the mass of displaced air shakes our bodies.

The highway is thick with traffic, thicker than yesterday. It's as if yesterday they weren't prepared to take the risk of trying to get through, and today there is a week's worth of backed-up cars and transport trucks on the highway. For the first half-hour of the day, the looming, speeding, roaring vehicles are a revelation, and almost an affront. What are they doing going this fast? By what right are they this loud? How can it be that such large, dangerous objects hurtle by with such abandon? Are these drivers unaware of the fumes they're spewing into our faces as they pass? Do they not know what they do?

But as the day progresses, we realise we are the aliens.

We are foreigners in a land of combustion engines. Ten thousand vehicles bear down on us from ahead, ten thousand from behind, and I am struck by how fragile we suddenly are – our soft flesh, our brittle bones, the infinitesimally slow pace at which we proceed through this barren landscape. How vulnerable.

Leaning into the sound and the smell and the roiling currents of air for hour after hour, I begin to see the highway as more than just the flat ribbon of dark bitumen laid out across the landscape. The highway is three-dimensional. We're not walking on it, or beside it – we're walking *through* it, and the roadbase and bitumen are just the floor of this thing. On either side of that floor rise its walls – the earthen embankments and fringes of eucalypts. It's hard not to compare it to the bed and banks of a river. Hard not to see ourselves as swimming and drowning in this hostile current of highway.

~

Many is the way suited for foot – created by it, or made for it. All the millennia of paths and tracks and byways of man and beast. Highways are not among that number. Highways are hostile to walkers.

~

The effort is draining. My pack grows heavy. Joints begin to ache. I suck water from my bladder and watch Ian do likewise. There is no breeze. Even so, the rain, which has been falling lightly for an hour, pulls off and the cloud hesitantly lifts. Encouraged, I stop to tuck my rain jacket away, take more water. The bitumen is buckling from the weight of a thousand trucks. A semitrailer pulling two silver fuel tanks roars past. I turn my back on it as I adjust my straps.

Uphill for a stretch, white rumble strip, grass growing thickly on gravel, *Eucalyptus tessellaris* saplings on the ridge above a cutting. I follow the highway's detritus as much as I follow the highway now. Flattened cane toads, a large python, bent grasses. So much refuse tossed from so many car windows – a catalogue of shame and satiation – cigarette packets, cans of Coke, a vape tube, beer bottles and beer cans. A box of sandalwood incense sticks, an iPhone recharging cord, occy straps.

Frogs croak in a marsh somewhere to the west. We turn our heads. Out on the highway, a frog is a creature from another realm, a mystery. We spy a split-rail stockyard but hear the frog no more. Perhaps we imagined it. We cross Ti-tree Gully. The gully is choked with weeds, but we can spot only one mature ti-tree, some distance upstream from the bridge.

Up ahead, another bridge and another creek, this one larger than the last.

We approach from the north. The bitumen narrows. There is no verge. *Logan Creek*, the sign reads as we near, and then below those words in smaller print, *Captain Logan Bridge*. We hurry, wary of being caught if a vehicle travelling at a hundred kilometres an hour reaches the bridge before we've made it across. Even so, I can't help but pause halfway across and look upstream until Logan Creek bends left and disappears. The creek is flooded, of course. Brown, and dirty with silt and refuse and slowly eddying leaves. Ti-trees and ironbarks up to their necks in water. I look downstream. That way is the swollen dam, Wivenhoe, out of sight behind a bend. Captain Logan Creek isn't flowing downstream though – the water under the bridge is backing up from the dam.

Logan Creek. Captain Logan Bridge. These are the first of a cluster of signs bearing that name, *Logan.* Later, there's *Logan Inlet Rd, Captain Logan Camp Ground* and *Logan Complex.* Yet nowhere in all of this signage is there a clue about who this Captain Logan was.

How do you read that silence? Is it laziness? Embarrassment? Discomfort? Is there something in all this signage with its repetitious name – the familiarity begetting forgetting – to be ashamed about?

~

Captain Patrick Logan died on the banks of this creek, downstream from this bridge, in October 1830. The precise location is now under the waters of the lake, submerged by the great drowning. But sometimes when a thing falls out of sight it doesn't disappear. It returns. Larger and more influential.

Logan has become myth. That is neither right nor wrong, merely inevitable. And probably necessary. Our lives are only ever partly our own. They're also palimpsests from which others find and make meaning. I wonder how much of Patrick Logan – born at Berwickshire on 30 November 1796, died 17 October 1830 on Jinibara or Jagera land beyond the boundary of Moreton Bay Penal Colony – still exists. Logan was a regimental soldier, a captain. He fought for what might be crudely described as a liberating army – the British – in Spain in 1813 as it freed the Iberian Peninsula from Napoleon's invading force. His was an iron discipline; the peninsula had been littered with broken armies and broken men. And after the Peninsula Wars he travelled: to North America for the war between the US and the British, to Paris in 1815 as part of the occupation once Napoleon fell, after that to Ireland and marriage. And then, when his regiment – the 57th – was ordered into Colonial Service in 1825, he travelled first to New South Wales and finally, the following year, to the new penal colony, Moreton Bay, as its leader.

Logan commanded the penal settlement with a ruthless brutality. So it is said.

He was killed hereabouts on the banks of this creek. Probably. Killed while surveying the river and its tributaries with a small party of soldiers and prisoners. Killed while riding his mare alone, yes, this austere and cruel loner. 'Killed by blacks', concluded the report to the commanding officer of the 57th Regiment. Killed by convicts, whispered the colony. Killed by both in concert, teased others.

Logan entered legend early, through the folk song 'The Convict's Lament' or 'Moreton Bay', written, perhaps, by convict-poet Francis McNamara. The song's narrator – from 'Erin's island but banished now to the fatal shore' – describes the most brutal of the penal settlements in a land teeming with them:

> For three long years I was beastly treated, heavy irons on my legs
>   I wore,
> My back from flogging it was lacerated, and often painted with
>   crimson gore,
> And many a lad from downright starvation lies mouldering
>   humbly beneath the clay,
> Where Captain Logan he had us mangled on his triangles at
>   Moreton Bay.

The Irishman exults at Logan's death:

> Fellow prisoners, be exhilarated, that all such monsters such a
>   death may find!
> And when from bondage we are liberated, our former sufferings
>   shall fade from mind.

Faded, perhaps, from the convict narrator's memory, but captured forever by song. Lines of which are echoed by Ned Kelly in his Jerilderie Letter. And wincingly acknowledged in Robert Hughes' *The Fatal Shore*.

Logan threatens to become myth. But killed, he was. A violent vision, a violent end.

~

We leave Logan Creek at midmorning. A black-shouldered kite hovers above us, pure white breast, fiercely beating wings, its head still as the earth. It knows something.

What we know, right now, is that the road is hard. We fall into an irredeemable slog. The noise, the displaced air of every car and every truck swelling against us. They are relentless, and the energy it takes to withstand them is immense. Step after slow, plodding step. Perhaps there is something in the carbon monoxide and the nitrogen oxide we're sucking in that is stupefying; perhaps the particulate in the exhaust fumes is already lining our lungs, making it harder to breathe. It's growing harder to think. Walking often leads me away from my thoughts – it's one of its blessings – but this is different.

By 12.30 pm, patches of blue have opened up above us. I've been trudging with my eyes to the tarmac for so long that I'd missed this shift in the weather. I stop when I realise. It is the first time I've seen blue in ten days and it comes as a shock. How easy it is to forget that change is possible, what it might look like.

The highway crests a ridge with views over the ploughed fields fed by a downriver tributary, Lockyer Creek, to the southwest. The road dips and rises again to Wivenhoe Hill and views to all points of the compass. There is an abattoir at the intersection between the highway and a connection road. We walk past. Having done so, we begin to observe fully laden cattle trucks rumbling towards us on their way to the meatworks, and passing us again going in the opposite direction, emptied.

Now the sun appears. I look up, then look away, blinking hard. We pause to drink. But there is nothing to celebrate in the sun's re-emergence after so long. The bitumen dries, the sun starts sucking moisture from the verge. From the leaves and the grass and the bush and the very earth. It's as if the heavens have abruptly changed their mind and want all that rain back. Immediately. Suddenly, it's hot and humid. Heat bounces off the highway, into our chests and into our faces. I hear the rasp of air as Ian sucks his water bladder dry. He'd only packed two litres this morning, and we still have four or five kilometres before we reach the dam wall. Ian's out, and I know I'm very close. On we trudge. Dry mouth and throat, head just starting to throb. We're dehydrating and ration what we've got left, taking turns to sip. There are puddles of water on the side of the road, and thousands of drivers to flag down if we get desperate. But we're very close and there's a picnic ground just this side of the dam wall, where there's sure to be drinking water.

~

Sometimes it's only experience that can transform information into knowledge. Perhaps even into truth. There are medical encyclopedias filled with the facts of homeostasis and the human body. The centrality of water. That water is sixty to eighty per cent of who we are. That we need it to digest food, to absorb nutrients, to excrete waste, to regulate body temperature. That every chemical reaction of the body takes place in water. That while we could go weeks without food – because our bodies have stores of fat and nutrients – we can only survive for three days without water.

That number – three days – is scored deep. During my father's last days, my mother and siblings and I sat by his bedside. He stopped drinking water, and a nurse or doctor or experienced friend said, 'three days'. But he was very weak. Would his body even last that long? And so we sat and spoke to him and prayed and wept and sometimes sang to him or read poetry, and his days without water passed. Three, four, five, a week, a damp cloth against my father's lips in ministration. Relief. Seven days and then relief, Amen.

So the body is made of water and, without it, falters.

Just how rapidly dehydration occurs depends on how quickly we lose water: through perspiration, water vapour when we exhale, urine when we take a leak. Ian and I have been pouring sweat, and – at times – breathing hard. Lose four per cent of your total body water, the data says, and you're dehydrating. Our blood's plasma loses water. Decreased water in the blood reduces blood pressure, the heart is forced to work harder. The kidneys have nothing to work with. The body's thirst response is triggered. All the body's water-reliant cells in all its water-needy organs become deprived and start to strain.

Thirst, overheating, dizziness, weakness. We're dehydrating, yes. We press on.

~

In the grass just off the verge we come across a spray of bones; a wallaby has been struck by a vehicle and thrown – or dragged itself – to its final place of estrangement. The bones have been picked clean and rearranged by dog and crow and insect. The puzzle of vertebrae, the delicately curved ribs, the twin scapulae. There are the large bones of its legs, and I wonder if, like mine, they are also called femurs. Not that how we make sense of the physiology of our fellow creatures makes a difference to the limitless, and unmourned, victims of road carnage. At the edge of this sacred circle – yes, perhaps I am sacramentalising it, atoning for my complicity – lies a white plastic bag, tossed and blown and weighted down now by catches of rainwater.

I am thirsty and all I can see is the death of those attempting to make passage through this land, a feverish panascope of roadkill and hunted deer and dead commandants and poisoned Kilcoy tribes and frontier blood and cattle trucked for slaughter and there are no words.

We stopped talking kilometres back. We are walking to our own rhythms. Another rise, and this time the highway passes through a deep cutting in the ridge line. To the left is a complex of thickly wooded gullies. The road ahead is a long, straight, dappled descent. The cutting seems to trap the traffic noise, magnify it. As the next semitrailer rumbles towards us we perform our ritual two-step – off the bitumen as it approaches, back on again when it passes. The truck crests the hill and disappears behind us. But strangely, a tremor of sound remains, seemingly from in front of us. Perhaps it's the lingering echo of the truck just passed, reverberating off the machine-cut walls of rock. Perhaps it's another truck in front, one we can't yet see. But there is a different texture to the roar ahead of us, and eventually, I look up.

At the end of the highway a huge cloud of mist is rising into the sky. It takes a moment to understand.

'Ian, Ian!'

I am struck with wonder, pointing dumbly, some form of paradise ahead of us.

~

It is a place of astonishment. I call Donna who drives from Esk to join us, to share this. The mist envelopes us all, soothes us, soaks us, elates us. We lift our heads to the sky, laugh and allow the mist to fall on our cheeks, tongues, eyelids. Ian and I have left our heavy highway selves a long, long way behind. Left the heat and the miles of barren bitumen to get here, below the dam wall at what's called the 'Spillway Common', to experience this – the release of floodwater at the spillway, to watch it course through the opened dam wall, inconceivable volumes with extraordinary power. The torrent bursts out of its narrow openings, hits a curved concrete ramp designed to disperse energy and propel the water up and out, into the province of sky, before crashing down again onto a series of concrete blocks and stones in the bed of the spillway, detonating into another cloud of mist, pouring and roiling and churning downriver in a foaming mass.

It's too loud to hear one other. We can barely even see each other. The afternoon sun pierces through the mist, arrow upon arrow of light. We squint to protect our eyes from the spray. Where does this mist end and the clouds in the sky begin?

But the three of us are not alone here at the common. Children squeal around us, slip over on the wet viewing platform, pick themselves up, lose their footing again. Their parents – up from Brisbane to show their kids this spectacle – abandon their efforts to keep their families dry. Cameras are exchanged. Delirious laughter. We drink this in together, agape, our blood rushing and coursing and exploding into something primal. We drip with water. We burst with its energy. We are cascade, we are rainbow.

~

We aim to camp at the confluence of Lockyer Creek and the river. Jason – councillor and veterinarian and supporter of efforts to restock the river with cod – and his wife, Laurisa, own the property.

'The ground is pretty bloody wet,' he warns.

We're used to wet ground.

'Make yourself at home then. Laurisa and I will be over in the morning to give you a tour of the place. The cattle might get curious, but they're nothing to worry about. And the shower in the farmhouse is yours if you need it.'

~

Donna drives us from the dam wall to the farm. It's only a couple of kilometres downriver from the spillway, but impossible to reach by foot in these conditions. Before the rains, before the flooding, before the need to manage the release of water, the dam operator had offered to escort Ian and me down from the dam wall to where we could rejoin the river. But the engineers have other things on their minds right now. And the usually dry Spring Creek, which enters the river from the west, is swollen and uncrossable for kilometres upstream. It's nearing dusk. Donna drives us back along the highway and then down along a series of jinking backroads till we reach the river once again. She saves us twenty-two kilometres of bitumen, and a day of road walking to advance a couple of kilometres downriver from the dam.

~

We pitch our tents just outside the home fence. In front of us is an oxbow lake marking an earlier course of the river. The cows are inquisitive, and as Jason had warned, the ground *is* boggy. I lay a plastic sheet under the tent. Flying ants begin to swarm, covering my yellow tent in black. A hare dashes across the paddock and under a hedge near the house. The sky is cloudless in the west, and the burnt-orange sunset through the door of my tent is achingly beautiful.

Day 16

Lockyer Creek to Twin Bridges

Distance: 17.02 km

Evening Camp: Place of the sheltering bridges

We rise to low cloud and a weakened sun. We rise to a low rumbling in the north. We climb out of our tents in the dawn mist and turn to the sound and realise what we're hearing is flood release pouring from the dam's spillway above us.

~

Dams are both beguiling and bold.

It's beguiling to think a dam might make us safe. How mistaken we are.

We thought, after the dam wall was erected in 1984, that it would prevent the city being flooded like it had been in 1974 and 1893 and 1841. But it didn't, and couldn't. It might take the edge off the damage a flood could do. Though in that there is a cost. Complacency.

Following the commission of inquiry after the 2011 floods, the commissioner described the dam wall as 'at once the most valuable and the most dangerous piece of infrastructure' in the state.

Valuable because the dam provides most of the city's drinking water. Dangerous because if it fails, hundreds of thousands of people face being swept away by the volume of water released.

That upriver roar is the sound of risk management. Of dam engineers releasing water that might cause some flood damage now, but not as much as if the waters are held any longer and the rain continues to fall.

But the roar is also the sound of dam engineers protecting the wall of their dam by releasing water now. Of dangerous infrastructure being preserved. Because there's another risk: of the dam itself collapsing. Because dam walls fail. All the time.

One of our earliest human attempts to construct a large-scale dam failed. As the pyramids were being erected 4500 years ago on the Nile, the Sadd el-Kafara Dam was being constructed across the narrowest stretch of the Garawi wadi in the desert to the east of the Nile River Valley. Its ruins tell us it failed even before it was completed. But dam failures are not a phenomenon of antiquity. In 1928 the St Francis Dam wall in the United States began to leak. The dam had been a grand project to store two years of water for Los Angeles, fed by the city's famed aqueduct. Leak became crack became collapse became 431 deaths. In 1935, the upper Orba Valley in Italy was rapidly inundated after a long drought when the Molare Dam was unable to cope with the pressure of the full reservoir. Swept away were the wall, hydroelectric plant, three villages, four bridges and 111 people. In 1959, the Malpasset Dam over the Reyran River in France reached official capacity and continued to fill as it continued to rain. Its entire wall collapsed, and 423 people died. In 1975, the Banqiao Dam in Henan, China, collapsed following cyclonic rains. Twenty-six thousand or more people died. In 1979, the Machchhu II Dam wall in Gujarat failed when it was simply unable to cope with the demands of its flooding river, and thousands died. The Kantale Dam in Sri Lanka in 1986, the Situ Gintung

Dam in Indonesia in 2009, the Ivanovo Dam in Bulgaria in 2012, the Panjshir Valley in Afghanistan in 2018.

And as the dam walls constructed in the golden era of mid-twentieth-century dam-building begin to age, so increases the risk.

So release the pressure of floodwaters on the dam wall. Release water. But get the timing wrong and what a dam operator releases is not merely water but a torrent of floodwater, furious at having been held back, intent on havoc. What is released is devastation with a human signature.

As occurred in 2005 when the waters of the Indira Sagar Dam on the Narmada River in India were released without warning to pilgrims bathing at a downstream sacred site. More than sixty people died. Or the discharge of flood waters from Hirakud Dam on the Mahanadi in 2011, when eighty people drowned. The release of dam water can be deadly. How deadly? How to warn? How to save?

Who would want to be a dam operator?

~

Jason and Laurisa arrive with takeaway coffees just before 7 am. Jason's a big man, generous. There's not enough room in the Polaris for more than three, so Laurisa, equally generous, waits in the car as Ian and I climb into the four-wheeler and make for the river. Jason stops at the high bank. The torrent of white water being released from the dam wall courses past in front of us. It's been pouring out since at least yesterday. Jason cuts the engine. We listen to the roar from the spillway. We listen to the thrash of foaming spate in front of us. The spectacle is still breathtaking: the width of the water, the flooded midstream red gums, the eddying foam, the different speeds of the currents – faster on the far side than nearer us – the unleashed power of it.

A cormorant dries its wings on a midstream ti-tree sapling, its trunk bent parallel to the river. The tree bounces up and down in the current, the cormorant riding it with outstretched wings.

'Now let me take you to Lockyer Creek,' Jason says.

~

Lockyer Creek. It's one of the largest tributaries of the river. It's over a hundred kilometres long, its own tributaries draining the range to the east of my childhood home in Toowoomba. I've explored those creeks as they ran down the range, played in them, dreamt in them. I've passed through the Lockyer Valley more times than I could count. Known its villages, drunk in its pubs, bought its produce, retold the stories of my father's and grandfather's comings and goings in the valley.

But more recently, Lockyer Creek has become synonymous with overuse, unsustainable agricultural practices, the leaching of nitrogen and phosphorus into the water, rising salt, extreme erosion. Too often the creek doesn't flow and becomes just a series of pools.

Perhaps you can't destroy water, but there are a thousand ways to kill a waterway. Deliberately or inadvertently. Slowly or overnight, and using different techniques. You can wipe a river from the face of the planet, or you can change its nature so fundamentally that it becomes a mere canal, soulless and barren.

If a city was going to take its time to kill a river, it would build industrial plants on its banks: boat-building operations, paint factories, foundries, chrome-plating businesses. It would pour effluent into it. Toss heavy metals in: mercury, copper, magnesium, calcium. Wait for the toxin levels to increase. Create a watery hellscape. Think relocating your heavy industry will solve things? Well, it helps. Mightily. But the thing about those chemicals is they settle in the mud, dormant, inert, just waiting for a flood to stir them up and set them loose.

Toxic chemicals aren't the only danger. A city might also suffocate a river with nutrients, that chameleon word. Don't think 'nutrient' is a healthy word. Overwhelm a river with nutrients and see what happens. Watch the levels of oxygen plummet. Watch the algae bloom. We might think for a moment how pretty those colours are, but then we'd be forced to think some more. Watch the waterways choke, watch those fish die, hear their death gurgle. Can we?

But which nutrients, and where do they come from? Name the culprits. Nitrogen and phosphate. Both occur naturally in soil, but also unnaturally, in the fertilisers we use to help our crops grow. And when those nutrients are washed by rains into creeks and rivers ...

Nitrogen: the most abundant element in the atmosphere, comprising nearly eighty per cent of the air we breathe. In a compound form plants use it to make amino acids that make the proteins for cell formation. And, critically, for chlorophyll. No chlorophyll, no photosynthesis – that great magic performed by plants as they use sunlight, water and carbon dioxide to create oxygen and glucose.

With increasingly intensive farming in the nineteenth and twentieth century, soils were losing nitrate more quickly than it was being resupplied naturally. Enter chemists Fritz Haber, who devised a way to combine atmospheric nitrogen with hydrogen to create liquid ammonia, and Carl Bosch, who then industrialised that process. Their synthetic nitrogen fertiliser gave birth to the Green Revolution and its extraordinary increases in crop yields. Which in turn feed most of the billions of us. But while nitrogen feeds us, if we're not careful, nitrogen pollution will kill us.

Phosphate: farmers have turned soil with phosphate forever. Manure, bones, shit. If soil is low in phosphate, you can't grow as many plants. So get phosphate from somewhere else – mine it, refine it, ship it, package it, transport it. Call it fertiliser. Add it to the soil.

Nitrogen. Phosphate. They're young words, born of the age of chemistry. Without the periodic table, would they even exist? As words, they are so much younger than 'land' and 'water'. Than words like 'farm' and 'crop', and 'reap' and 'sow'. Than harvest moon and summer heat. Than feast and famine and plague and pestilence. It's hard to weigh young words like 'nitrogen' and 'phosphate'. To know whether they've got heft. Whether to respect them or to fear them. Whether or not to take them seriously. We should. We have no choice.

~

Jason, Ian and I stand at the junction of creek and river. We see soil and sediment being swept into the river. The rain and run-off sweep leaves into the great stream, strips of bark, sap-drip. Leaf litter – blue gum and she-oak and callistemon – freshly fallen or decaying is washed into the water. The soil below the carpet of leaf has been made rich with dissolved organic matter. Intense pulses of chemicals and nutrients – both organic and manufactured – are swept and leached in the river and flow, flow, flow.

All the sediment that washes out of the Lockyer Valley down the creek and into the river and all the way through the city and into the bay. Can there really be a saltwater bay at the end of this? It's hard, in this moment, to feel the connection. Hard to conceive that it may take sediment seven slow years to meander from the valley to the bay. Or forty-eight hours in a flood. Assuming, that is, your starting point is the dam wall. Who knows what is gathering on the upstream side of the wall?

We three stand at the mouth of Lockyer Creek. The creek is flowing much more slowly than the river. At least, Jason says, it is today. The dam operators can control the speed of the river or the creek, not both at the same time. Jason explains that the dam operator had delayed the release of floodwater from the dam to allow floodwater in the Lockyer Creek catchment to come down first. But holding the water in the river back means the floodwater rushes down the Lockyer more quickly than it naturally would, because there is not as much water in the river ahead to slow it down.

A consequence of this faster flowing water? A great, unnatural, erosive force.

I stand at the confluence of the rushing river and the stilled creek. It's as if the creek has stopped to give way to the traffic of swirling foam coursing down the river. Upstream, the banks of the creek gape where land has fallen into water. On the far bank of the creek the erosion has reached the very edge of the cultivation, gnawing, chomping. A power pole is centimetres away from falling into the water. On this side a magnificent blue gum holds the bank where upstream and downstream the bank has fallen away. But it feels like a last stand: glorious and noble, but doomed to fail. As much as it would hurt him, Jason says he'd be prepared to lose the blue gum now in order to batten the creek bank properly in the hope of saving it.

People have an extraordinary capacity to forgive. But nature? I look at the disappearing bank. Think of the Brisbane River cod and the Tasmanian emu. Never. The offences we commit against rivers and forests and the air remain eternally unforgiven. It is for us to atone, if it is not too late.

~

After Jason and Laurisa leave, Ian and I finish packing. For Ian it's for good, which means this journey will change forever, once again. Ian has to get back to his family, his work. It's not long before Alisa arrives to collect Ian and drop off Grantley – an old schoolmate who's become a hydrologist – and to resupply. We all wander down to the creek again, this time to the weir.

O'Reilly's Weir is a kilometre upstream from Lockyer Creek's confluence with the river. In drier weather there are any number of places to cross by foot what's often a sad and spluttering waterway below the weir. It might even be theoretically possible to walk across the weir itself, tiptoeing along the top of its narrow concrete lip from one side of the creek to the other, careful not to overbalance and slide down the long concrete face. But that would be foolhardy, an escapade of teenagers proving themselves, not of responsible men with heavy packs. On other days you might even swim, floating your pack across, or persuade a farmer like Jason to row you to the other side. But not in these conditions, not today. The weir has disappeared beneath floodwater. *DANGER. WEIR. STRONG CURRENTS AND UNDERWATER HAZARDS CAN BE PRESENT.*

We climb into Alisa's car and drive the 12.5-kilometre detour which takes us across the creek further upstream at the Patrick Estate Road bridge and leads us back to the southern side of O'Reilly's Weir, where we read its twin warning sign, *DANGER*, once again.

We're now in the heartland of Lockyer Valley produce-growing country. A local farmer approaches us on his quadbike: long-sleeved blue workshirt, dark blue shorts, floppy hat and barefoot. He points out the erosion of the banks from this side of the creek. We stand as close as we dare, taking our cue from him. Too close and our weight will dislodge more earth and we'll slide into the water. It's a seven- or eight-metre fall where the February floodwater gouged out metres of land. The floods this week have taken more, leaving freshly bared earth, which the water is now quietly teasing away, undermining it lick by gentle lick.

White plastic piping protrudes uselessly from the bank, or hangs, snapped. Orange conduit pipes and electrical wires worm through the turned earth, whatever original purpose they had now broken.

~

Ian and I must farewell each other. This is my third farewell. I hadn't foreseen how unsettling they would be, the sharpness of them, only partly tempered by being joined by a new companion. I think of the poet Basho farewelling his walking companion, Sora, on his journey along his narrow road to the deep north of Edo Japan in the late seventeenth century: 'the sadness of the one who goes and the grief of the one who is left behind'. Though there are different qualities to these aches of the heart, shaped by the nature of the relationship: new friend, son, and now brother. How much can a brotherly embrace on the banks of a tributary hold?

~

The way between the irrigated paddock and the collapsing bank is a strip of mud. Our boots are soon thick with it, growing larger and heavier with each clinging clump of clay. Grantley was once a nationally ranked junior athlete and still moves like one, with an indefinable mix of grace and strength. But the mud gets the better of him too. The tread on the soles of our boots disappears beneath layers of muck. With each step we slide, mud on mud. Or sink, before having to tug our heavy boots out of the mire, leaning on our walking sticks, which in turn sink into the depths and need to be pulled free. We'd be better off barefoot.

We track Lockyer Creek down to the confluence and pause a moment for Grantley to take it in. I press him, flood expert, for his first professional reaction to it all: the stilled creek, the velocity of the flood release down the river, the high-water mark from February, the water level now, some unusual eddying. He winks. 'It's pretty foamy.'

We follow the corrugations of tractor tyres in the mud, the rushing water to our left, the prime produce-growing paddocks of the Lockyer Valley to our right. Some fields are fallow, others teem with young crops. Including – of all the crops to grow in this country of such variable rainfall – rice. Another farmer, barefoot like the first, spots us and rides over on his four-wheeler to ensure all is well. His farm has been here for generations, drawing water for irrigation, growing the full palette of vegetables: greens and browns and reds and yellows. We talk about floods and the economy and the upcoming federal election and crazy pilgrimages along rivers.

'How well you must know this river,' I prompt.

'Bastard of a thing,' he replies and waves us on our way.

February's flooding is still fresh, and who yet knows if this rain dump is finished.

This side of the dam, the evidence of February is clear. Though the river is up now, coursing and powerful, the water level is nothing compared to February's floodmark. The bank we're following must be ten metres above the surface of the river, and in February the water topped the bank. The barbed wire fence we're walking beside is still caked with dried grass caught in its strands as water rushed through. Attempts to burn the debris off since February – that time-honoured way of cleaning fences after flood – have failed. The grass is too thick and is merely scarred black around the edges. Metres above our heads, the branches of river gums are festooned with black packing tape. A car tyre is suspended over the river, impaled on the overhanging branch of a dead tree like a single quoit thrown perfectly onto its spike.

We leave the riverbank to pass through the village of Lowood. Outside the Vietnamese bakery on the main street we unhitch our packs, buy fresh sandwiches and collapse at one of the silver tables on the footpath. People come and go – tradies, parents with kids, local business owners. Grantley has come up from Sydney for the walk, and the lunchbreak is the first chance we've had to check in on each other, and share stories about our families and mutual friends. Time passes more quickly than we realise, both this lunchtime hour and our lives.

When it's time to get going again, we cross the road to a pretty tiled mosaic of the river on a monument in the main intersection, with the story of the village told on its plinth. The history begins with 'the traditional owners of this country who lived here for countless generations' and ends with an explanation that 'Lowood is now a place where many people choose to live for its ideal lifestyle – a relaxed country town near a picturesque bend of the Brisbane River'. The real estate agency advertises properties for sale in its street window. An Aboriginal flag is draped from the window of a cottage on the route out of town. A truck driver who'd greeted us back at the bakery honks his horn as he passes.

Up ahead is Lowood Bend, an unofficial picnic area, a sanctuary for young lovers. It is here in the late afternoons, no matter the season, that people from up and down the valley come in their four-wheel drives and find their own secluded beach, lay out mats or start campfires and lean back with bottles of beer or bourbon and lose themselves while the river riffles by. Each time I visit here, the refrain from Leonard Cohen's song comes to me, as if it's Suzanne taking me by the hand, taking me down by the river.

Because what lover has not taken their beloved by the hand and led them down to the river? Who has not made love by flowing waters, seeking and finding there everything and nothing in equal measure? If we are not Suzanne, we've been with her as she takes us to her place near the river, close enough to hear the boats go by, and to spend the night with her, knowing she's half-crazy. We lose our sense here, though not our senses. In these moments the river sweeps away reason and offers something else in its place.

~

The picnic area at the bend of the river is, of course, flooded. Six months earlier, a scientist friend, Daniel, and I had taken water samples here. It was one of a number of places along the river where we'd gathered water to test, in truth, what was more a novelist's hypothesis than a scientist's: might the micro-organisms in the water at different sites along the course of the river reveal something about the river's character as it flows from its source to

175

the saltwater bay in the east? (Answer: micro-organisms especially love stagnant pools of river water isolated from the river as the floods recede.) This particular beach has disappeared. All the riverside walking paths and four-wheel drive tracks – and therefore our planned route for the next leg of the walk – are also submerged. We'll need to stay higher, and cobble together an alternative route.

The colony of flying foxes at the bend is agitated. The animals leave their roosts, then return almost immediately to the same place on the same branch, screeching. The tree shudders, unsettled. Patches of high blue sky, scudding clouds, dark twisting bodies and translucent wings. The flying foxes and their river roil and start.

Adjacent to the flying fox colony is what looks like a new water station, loaded with electronic equipment. Grantley gives me a lesson on telemetry as we pass. 'The sensor and gauge measure the water pressure and the water level, and the data gets transmitted in real time.' He points out the various instruments and the role of each. We record so much about the river and think we know it.

The afternoon follows the river. An afternoon away from the flooded bed and bank, but keeping the water in sight. We walk along the rail trail, or along the verge of the road as it hugs the river, or through riverside paddocks. It's an agitated afternoon, the mood set by the disturbed flying foxes. A Charolais calf is separated from its mother by a barbed wire fence. An old besser-block shed on a small concrete slab bears a skull and crossbones symbol in a triangle beneath a lone word, *CHLORINE*. High-fenced properties with concrete gargoyles and snarling dogs fringe a riverside road of small allotments. But then we come across a dreamer manufacturing canoes from his riverside property. I'm grateful for his dream, for the picture in my imagination of handcrafted canoes pushing off the bank in the falling light to follow a gentler destiny.

~

It's dusk by the time we reach the Twin Bridges Reserve campsite, named for the two river crossings here, each spearing off in a different direction. We'll be sharing the site tonight. On a rise to the left is a semicircle of three large tents facing a portable table, fold-out chairs and stove. The table is on its side and tea towels hang over the backs of the chairs or lie, windblown, on the ground. The door-flap of one of the tents is hitched open, revealing mattress and doona. Plastic bags, water bottles and a cardboard box lie strewn around the common area, weighted down by rain. But we can't see anyone. Nearer the river are two covered picnic tables set on concrete slabs. An old model Nissan Patrol is parked beside one, its bonnet up and back door open. A man, perhaps in his late fifties, sits at the table dipping corn chips into a jar of salsa and pouring soft drink into a blue plastic cup. Another vehicle, a white HiAce van, is parked a respectable distance from the Patrol.

Grantley and I unshoulder our packs at the second table. I call out a greeting to the man, who rises. Dark blue t-shirt, dark shorts, white socks and KT-26 sandshoes, sun-splotches on his legs, slicked-back hair.

'Are you marking the road?' he asks when he reaches us, taking us, in our boots, for council workers.

'No, walking the river.'

'Oh.'

We all look out to where the sun has fallen behind the horizon. There's beauty out there.

'I'm keeping an eye on the bridge,' the man says. 'It's still under.' He is standing erect, like a captain on duty. 'Can you see?' he asks.

I follow his pointing finger. Across the swollen river, Wivenhoe Pocket Road dips down to the water and disappears. There's a vehicle parked beside the road on the other side, and beside it a solitary figure in a camping chair.

'I know him,' he says. And then, nodding towards the HiAce, 'I know him too.'

He welcomes us – the solidarity of fellow travellers – and shares what he knows. Offers it willingly, freely. There is a toilet block up the end of the track near the highway that gets locked at 6.00 pm. We've got twenty minutes left. We may not notice it at first, but there's a tap positioned high above the urinal in the men's which is the best tap for filling water bottles. As for the tents over there – he gestures – the owners abandoned them ten days ago.

We fill our water at the high tap in the toilet block and pitch our tents in the dark. For the first time this trip, mosquitoes swarm around me. At the picnic table, under the light of head torches, we smear insect repellent on our exposed skin. We prepare our meals and offer whiskey to our companion. Behind us the side door of the HiAce slides open. I turn at the sound. An old man swings his thin legs down from the mattress laid out in the back of the van. He stands beside his vehicle for a long while, gazing out over the river.

The moon is just past full and rising steeply. It's been a long time since I've seen it. Grantley and I retire to our tents, the men to their vehicles. But the moon is too bright.

~

I lie in my tent and can't help but wonder what would cause a group of people to abandon their tents and camping gear – thousands of dollars' worth. Whether this reserve had suddenly no longer been a refuge for them, and they'd had to drop what they were doing and flee without time to pack. Whether they'd been taken away, voluntarily or against their will. Whether, while away on an errand – buying supplies in town, visiting people, having a drink at the pub – some disaster befell them. Or whether it was the rains that drove them away, before cutting the river and stranding them on the other side.

Though the too-bright moon keeps me awake, still there is the sound of the running river. Constant. The solace of flowing water.

If I sleep at all, it is because the river allows it.

Day 17

Twin Bridges to Fernvale

Distance: 13.14 km

Evening Camp: Place of the grass-tree forest

I want to slow down and observe. I want to have the time to contemplate what I see. To record carefully in my journal or to dictate the details of the world around me. The relationship between bird and water, light and water, bootfall and water. But I am overwhelmed. By the need to stay dry, to find shelter, choose a route, step safely. Demands that are more immediate than the luxury of meditation.

The river won't let me contemplate it. It is, instead, forcing me to reckon with it. And it is tiring.

~

One of the twin bridges is now open, but the other remains flooded.

Today's route takes us from the campsite, along the right bank, over the Brisbane Valley Highway where it crosses the river, to Zanows' Quarry. From Zanows' we'll follow the embankment between the quarry site and the river, and then, once we've cleared the quarry, between the river and Poole Mountain, across Ferny

Gully, till we arrive at a property belonging to local horse and cattle people, Andrew and Jill.

Except, as we discover when we get across the highway, the quarry has been flooded and the path along the embankment has been cut where the river has broken through. Perhaps we can walk through the quarry site itself and pick up the river on the other side? Standing at the front grid of the quarry, I ring the site manager, Dave, on the number shown on the sign, all visitors being required to report to the site office.

The quarry is enormous. I'd visited the year before and discussed its operations with Darren, one of the owners, though he and his brother have now sold it. All the way down the river valley, people have been whispering rumours of the sale price. We wait in the shadows of two flagpoles either side of the grid, the Australian flag and the purple company banner with its lightning-bolt Z. In the distance are pits and ponds and silos and mounds of different coloured earth and sheds and removable dongas and plant equipment and a fleet of trucks rumbling into life. Dave drives down to meet us.

'Yes, that's right,' he says, 'the path along the river is under.'

'Can we cut through?'

'Nah. Half the quarry's underwater. There's no safe route through.'

Dave and I look at each other. Dave would like to help, but there's nothing he can do.

'What if …?'

'No.' He shakes his head firmly.

~

'So, Plan B?' Grantley asks.

If I've led Grantley to believe Plan A is much more than just following the river where it wills, it's because you can't expect an engineer to leave his important work on carbon capture in mangrove ecosystems and fly up from Sydney for a loosely organised river pilgrimage.

Plan B, as I'm thinking it through, involves the property adjacent to the quarry, which is owned by a river family, Noel and Helen, and their son Ken. They've been there four generations in all, the first generation arriving from Germany in the late 1800s.

'The river is everything,' Noel said when we met on an earlier occasion over a cup of tea and baked biscuits. It's a character in every yarn, he says, with lungfish the size of canoes, and the river in flood so loud you could lose your hearing.

Father and son both attended the Fernvale primary school opposite their farm. There were sixty kids at the school when Ken was kicking around in the seventies, but that became 360 almost overnight when construction started on the dam. 'What changes the dam brought,' Ken reflected. To the river, of course, but to the community too – demographic changes, with labourers arriving, people from different cultural backgrounds.

Since the dam, the water always runs. 'It always did,' Noel hastened to add, 'but now its flow is more predictable. It changes about half-a-metre over the course of a usual week. Up down, up down.'

River people like Noel and Ken measure their hours by the river's height. Every glimpse, every peripheral scan, every conversation and scudding cloud are absorbed into their river-consciousness. It's as if their inner workings are set to the rise and fall of the river and respond to its most minute stirrings.

We had talked, that first time we met, about natural floods and man-made floods. Before the dam went in, the floods were natural. Since it was built in '84, the floods are man-made. January 2011 left its mark.

'The noise was incredible. The flood came out of the Lockyer Creek first. Water was pouring off the western flank of the D'Aguilar

Range. The force of the water coming down was so great it changed the shape of the river. You looked out and you could see a bow in the middle of the river. True. The water was higher in the centre of the river, so high that the bank on the other side was obscured by this meniscus. And then …' He pauses. 'Then it got so that there was so much water the river disappeared.'

~

Now, I ring Noel's mobile and hear someone fumbling to turn the phone to speaker-mode. He and Ken are in the car; they apologise for not being home. They're happy for us to walk through their farm to rejoin the river. Of course there'll be a route along the bank on the riverside flank of Poole Mountain. If they'd been in, they would have offered us a coffee to set us on our way.

The route Noel and Ken instruct us to take is a long, unsealed farm road. The generations of old machinery gathered on either side of the road offer a lesson in the evolution of farming practices in the district. We cross grids with water pooling in the hollows beneath. We zigzag around deep puddles and stop to offer a carrot to a chestnut mare with striking black socks.

Poole Mountain rises higher and steeper as we approach it. Cattle pads run around its base, appearing, at first, like a series of floodmark rings – running roughly parallel, but at different heights – around the mountain. After choosing a well-trodden clockwise pad, it soon becomes apparent that the tracks don't run parallel at all, but cross and merge and pull apart according to the contours of the land and bovine behavioural patterns we can't decipher.

Eventually the river comes into view, brown and wide and full of energy. We stop to catch our breath, suck water from our bladders, gobble down a muesli bar each and then a second.

We look back upriver at the quarry and the series of lagoons threatening to encircle it. The whole operation down there appears, from this vantage, so fragile, so temporary, no match for the river. All the demountable office sheds and trucks and silos remind me of a circus – easily disassembled, packed away and relocated. Because that's the nature of extraction businesses: they're inherently transient. That which can be mined is mined. That which can be crushed is crushed. That which can be separated is separated. That for which there is a market is sold, and that for which there is not is spoil. And once the earth is no longer of use, once it is reduced to spoil, the show moves on. As quickly, often, as it can.

~

The mountainside grows steep, falling away to the river on our left. The cattle pads narrow. Sometimes we lose them and have to guess our way across the high-grassed slope, tracking between lantana bushes and exposed rocks, the gums too sparse to provide shade. We sweat beneath the high sun and lean on our walking sticks, or shed our packs at intervals, to rest. The river hugs the mountainside. It travels quickly, leaves us behind, thinks nothing of us. We have gullies and logs to cross, thoughts of snakes to dismiss. We swing ourselves around the trunk of a young small-leafed fig. The sun sucks water out of the saturated earth. We sweat it back in large drops of perspiration. A wallaby crashes through the scrub high above us, thud, thud, and is gone.

Grantley is looking for a chance to talk. I'm not sure if it's because he's worried or if he just wants to call BS on my assurances about Plan B. Either way, I stay far enough ahead of him so he can see where to follow, but not so close as to allow conversation. When we stop to rest, I steer the talk to how completely confident I am about where we are and how I've been up here plenty of times before on planning trips.

After an hour, we find that the mountain is behind us, and we discover we're on a descending ridge that flattens and widens as it drops. Too late we realise it's a small peninsula with steep gullies both left and right. We look to see if there's a way across the gully to our right – heading more or less downriver – but the vegetation is thick and the water in the creek deep, so there's no option but to backtrack until we find a place to cross. Sure enough, there's a way. Once across, we climb the other side of the gully. Another surprise: we've arrived at the back of a residential estate. Dogs begin to bark, one after another, a snarl of guardians protecting their homes from intruders emerging from the bush. Behind the rear boundary fences of these newly built houses is a track running parallel to the gully; soon enough the river is before us once again, the dogs cease their barking, and we find we've arrived at a riverside park. The new housing estate sprawls away from the river up the hill.

We stop at a roofed picnic shelter. We need to rest, we need to eat, we need to drink. We're nearly out of water. A young home owner working on a double-tiered grey-block retaining wall is happy for us to fill up from the tap in his front yard. He's only been down to the river once, he tells us, a couple of months ago with his daughter – who's jumping now on a trampoline beside the house – and he isn't sure if there's a track along the bank. I spread triangles of La Vache Qui Rit cheese over pieces of pita bread, sprinkle them with pumpkin seeds and almonds and sultanas and then roll them into lunch, one for each of us. We take off our gaiters and boots and socks, and lie back on benches and close our eyes.

I wake moments later to the sound of crows' claws on tin above my head. It's an omen, I think – a good one. Since a trek along the Kumano Kodo pilgrimage route in Japan with my family a few years ago, I've seen crows differently, not as cursed or malevolent or indifferent, but as companion birds. Guides. A three-legged crow, Yatagarasu, had escorted Emperor Jimmu through the Kumano mountains after he'd become lost, and crows have, since then, been sacred.

Well, crow, where to from here?

The route ahead requires us to cross Ferny Gully and follow the river beyond Savages Crossing bridge – named after an early farmer of the district – for another four or five kilometres to Andrew and Jill's cattle property south of Fernvale township.

In drier times, there are walking tracks along the bank in front of this parkland that lead to Ferny Gully. In drier times, Ferny Gully can be walked, leapt or waded across. When I was here to reconnoitre last year, it was a drier time. Those riverside tracks are submerged now, and the way to Ferny Gully requires us to wade through waist-high castor-oil plants. And the water in Ferny Gully, when we reach it, is thirty metres wide where it joins the river.

Grantley knows better than most the risks of crossing flooded waterways. As a hydrologist, he has modelled the fates of various types of cars attempting to cross flooded bridges. He's dragged hatchbacks, sedans and four-wheel drives with their high ground clearance through a specially constructed flow-pool and then monitored the water level rise and calculated the point at which the vehicle begins to move, and then how long before it floats and is carried away. It's research interwoven with a thousand public education campaigns: *If it's flooded, forget it* and *If it's up, it's not on.* Ferny Gully is flooded, and we forget it.

We look for a route upstream along the bank of the creek, but from here near the mouth as far as we can see it is steep and dishearteningly impenetrable. We backtrack to the picnic area and follow the road as it runs parallel to the creek for a hundred metres before we plunge into the bush again in search of a crossing. But the vegetation is thick, thick, and as much as we bash, bash, our progress is slow and exhausting. We're losing time, and when we catch sight of the creek again it's at the foot of a steep drop and is dark and rank and forbidding. How apt that an earlier name for the creek was 'Stinking Gully'. But it is still too wide.

We look at each other. It might take until dark to pick our way through this suffocating forest before we find a place to cross. If there is such a place. And the other side looks just like this. Our satellite map offers no cause for optimism either. We decide to retreat to the table where we'd had lunch to work through our options.

No sooner do we collapse back at the lunch table than friend Crow alights on the roof above us. Is there a pattern to his scratchings? Some language in his movements waiting to be deciphered?

A lawnmower starts up in the backyard of one of the houses in the estate. If Ferny Gully can't be traversed by foot down here near the river, the options seem limited, and all involve different versions of us making our way by road out of this housing estate to the first vehicle crossing up on the highway in Fernvale, and then coming back down to the river by road on the other side. Fourteen kilometres by satellite map. The bitumen street leading up the hill is steep. Andrew and Jill are expecting us, and we're keen to get to their place tonight if we can, but if we have to detour by fourteen kilometres at this hour, given what's still ahead of us, we won't make it. Desperate times, desperate measures. I decide I will interrupt the gardener from his mowing and beg him for a lift into Fernvale.

But before rising from the bench, I pause, hoping for some sign from Crow.

Almost immediately my phone rings. It's Andrew. Where are we? Andrew is a tough nut. A superb and unstintingly generous nut, but tough. He's been guide and supporter throughout the planning of this walk. I've tested ideas with him. He's proposed others. How is it possible that he's calling at precisely this moment? What message must have passed from one guide to another? Of course, there's another explanation – that Andrew is organised, is expecting us, knows the river and is curious about our progress. If we stay put, he says, he'll collect us and drive us around the obstacle Ferny Gully has become.

~

When he unloads us at Savages Crossing – which is still cut and has barriers erected at the end of the road to prevent vehicles from blithely driving into the floodwater – he gives us directions along the riverbank to his place. We leave our packs in the back of his truck and set out on this last leg, lighter of step and spirit.

The route Andrew described is an itinerary of tracks and gates and paddocks and fig trees on ridge lines. High clouds gather and break. The afternoon sun comes and goes. The blue sky is pierced clean through by the dark silhouette of a kite.

The final approach to Andrew and Jill's home is a winding ascent by cattle pad past a stand of ancient grass trees, three metres and more tall. Has anyone ever walked through a forest of *Xanthorrhoea* and not been humbled? You know they're kindred. But older and wiser. You have to stop. Listen. Hear them sigh. Hear them murmur. They have borne witness.

~

Andrew and Jill's home on a ridge is new, with views both up and down the river. They are yet to finish the landscaping, but they've started with grass trees. They know to do that. The rest will follow.

And what follows a hard day's walking is sitting on a deck with friends and looking down at the river and its valley. What follows is sharing stories of the walk with Andrew, what the country was like, who we've met, how many kilometres we've made each day, whether the GPS has been useful. We sit on the deck and put up our feet and drink beer and pull cobblers' pegs from our socks. We sit and accept whatever comes tonight: teasing over our dirty clothes, the chance to put on a load of washing, a shared meal, camaraderie, the sweet drift of conversation threaded through by river.

This peninsula is full of stories. Andrew and Jill know a few. Andrew opens a bottle of red. Yarns morph into stories morph into legends, then fork off: fable this way, myth that. Local yarns with the meter of a bush ballad. But there are stories neither Andrew nor Jill has authority to tell. Speak to so and so, they say. Have a cuppa with someone else. Did you ask John on the property next door about Dan Kelly?

~

I had, on an earlier trip. Dan Kelly being bushranger Ned Kelly's brother.

John is in his eighties now, but he was a kid of five when the old man who called himself Dan Kelly lived in a hut at the foot of his property in the 1940s. John remembered him well, used to get lifted onto his knee, was frightened of him and his beard. John showed me an old sepia photo of the hut, a makeshift collection of scavenged timber and sheets of corrugated iron. I turned it over: *Dan Kelly's Hut – Harris Road, Fernvale. Approximately 1950*. The hut is long gone. So too Old Dan. Kelly had gone by the name James Ryan until age and guilt caught up with him. As he told the *Truth* newspaper in 1933, no, he hadn't died in the Glenrowan Hotel. His brother Ned had been captured, of course – everyone knows that – but *he'd* escaped the flames and made his way to Queensland. The two burnt bodies pulled out of the cinders of the pub and buried in the cemetery in Greta? A couple of friendless drunks. The old man would take off his shirt for anyone who cared to see the burns on his back.

The line between legend and myth can sometimes be difficult to discern. It may take generations to understand what a story tells about a people, a place. And how much does it matter whether itinerant James Ryan was Dan Kelly? Because the river is a snake, shedding skins.

Day 18

Fernvale to Fairney View

Distance: 11.79 km

Evening Camp: Place where the full moon burns the water

I am in cloud now. I am closer, out here on the deck, to the moisture in the sky than to the river's coursing waters below. The morning breaks slowly. The sun, when it arrives, is formless, its light diffuse. I am not yet ready for the day ahead; there are still more cobblers' pegs to unpick and it's good to be out here alone, my fingers at work on a simple task.

I finish and look up. In the sky, a spear of gold pierces the cloud.

Jill joins me for a coffee. We sit and talk. We listen as the day begins to hum. In time we rise. On a farm there are always chores to be done. As for me, I have blisters to dress and a backpack to fill.

The four of us gather again for a breakfast fry-up of eggs and mushrooms with toast, chilli jam and coffee, and to map out the day ahead. Having been chastened by Ferny Gully, Grantley and I are concerned about today's creek, the formidably named Black Snake Creek. On a reconnaissance trip last year, one of Andrew and Jill's neighbours, Russell, a retired accountant and farmer, had looked at me squarely with a challenge.

'How will you cross Black Snake Creek?'

I'd been blasé.

'We'll just wade across.'

It had been dry. The satellite map images had also been taken in the dry. Wading was the obvious answer.

'And the snakes?'

Back then, the answers to Russell's questions didn't seem to matter. Now they're important. Very important. I press Andrew. He knows well the large horseshoe peninsula he and Jill share with another couple of property owners. There are a number of crossings over Black Snake, he says.

'Where?' Jill asks him.

He lists them, including two private causeways.

'And if they're all impossible, there's always the highway,' he says, and he laughs. Because he's of the school that believes the harder something is, the better it is to have done it.

~

When it's time to leave, we take a high route out towards the end of the peninsula along a series of ridges, the river to our left. It's overcast, but not raining. Step by step, it becomes increasingly humid, the sun muted but powerful enough still to be sucking weeks of moisture from the saturated ground. We pass a swing with a river view hanging from the bough of a *Ficus obliqua*. We track a well-maintained electric boundary fence, scale it at the corner post, pass a rusting Southern Cross windmill and an old farm complex. A flowering of pale yellow fungi, delicately membraned, rings a collapsed termite nest. As we descend to lower ground, masked lapwings rise from the grass, urgent wingbeat and low trajectory. But it is not we who have disturbed them. It is a bone-deep rumbling that the plovers heard long before we did. How much we don't see. And how much we only see when it's too late. Eventually we look up. A military cargo plane, a monstrous C-17 Globemaster, is moving slowly across the sky. So slowly it is unnatural. Surely this huge object will fall. Surely there is some law – physical or moral or both – that will cause this thing to drop from the heavens, because whatever it is doing as it creeps across the sky, it is not flight. I follow its alien path and feel suddenly small, vulnerable, unsettled. But why should I? This is *my* river, *my* country. My tread is soft, my demands few, I walk the river in peace, seeking companionship and understanding. I seek no mastery over the globe, make no claims of that sort. Why should I shrink?

And then it is gone, returned to the nearby air force base, and as I continue to look, I see the cloud is fragmenting above me, and there are patches of blue. Is it possible that even the clouds have made way for that foreign beast?

~

The river pulls us hard towards it. We reach the end of the horseshoe peninsula and come off the ridge. Grantley explains what we're seeing as we descend to the water line. He starts with stones.

'See how they're becoming smoother as we drop to the flat?'

I bend and gather as we go, palm-sized stones, increasingly water-caressed.

'See too the different layering of sediment?'

I see sand and I see dirt.

'Remember sand is just ground stone,' he says patiently, 'and soil is organic. Watch the river carry the soil. Watch it carry silt. See how thick with soil the river is. This is where "fall velocities" come in. The principles are the same whether an object is falling in a vacuum, air or river water. The theory is easy – you can express the relationship between gravity, friction and buoyancy in a neat formula. The problem – and it's a major problem – is reality. The reality of how big your particle of earth or sand is, what shape your particle is, its density, how dense the water, how viscous the water.

To say nothing of what the currents are doing.'

We look out at the rushing, eddying, churning river. Grantley has spent years plumbing the complexities of fluid dynamics.

'But as a rule of thumb,' he says, 'sand is heavier than silt, and falls out of floodwater first.'

I look. I see.

That sandy beach, and that one. The one we're standing upon now.

The silt? It's on its way to Moreton Bay.

~

We shuck off our packs. We're within the river's magnetic field. As if we're grains of sand fallen from the river's universe. It's good to feel its energy like this, the kinetic draw of the surging river in the blood. On the far side is a steep cliff, rising sheer out of the river. About halfway across is a long line of trees that mark the usual boundary of the river, their trunks rising out of the floodwaters. Between those trees and the rock face is a deep channel. That's where the faster currents are, the floodwater coursing in a swift, rugged rush against the rocky face. The river has tens of thousands of years of work ahead of it over there, scouring the cliff, turning rock into sand. Kookaburras call out from the other side of the bank, two, maybe three. An intermediate egret lopes downriver.

We bend and examine and fossick.

I wander down to one of a number of natural debris traps – large piles of sticks and twigs and logs that have gathered at bites in the river. I pick my way around this stack of tree branches. These trees have done their work, I think – absorbing carbon dioxide, sequestering carbon from the atmosphere, photosynthesising, drawing water from the ground and transpiring it into the air – and, their work done, they collapse into the water and float to a downstream shore – here – to decompose. The pile of sticks reminds me of the funeral pyres I've seen on the banks of the

Ganges at Benares. In Benares, the pyres are constructed by human hands, for bodies to be cremated beside the sacred river and their ashes scattered into its waters, helping to free the deceased's soul from the cycle of death and rebirth. Here there's no body and no fire, and the river isn't breaking the cycle of life and death but continuing it. Work, I think, that is also sacred.

~

We follow the arc of the horseshoe, in time climbing up from the water to a cattle pad along the high bank. The path skirts a patch of remnant rainforest. An enormous banyan is centrepiece of the tiny forest, and it is irresistible. Inside, the little universe around the fig is dark and cool and quiet. The river, so close, momentarily recedes. Above our heads, a canopy of new leaf and ancient branch. Beneath our feet, leaf litter. Breathe. Respiration and transpiration and expiration. Living and dying. There is no end to the ways one can be made small.

As we come out of the forest there are cattle on the bank ahead of us, a herd of fifty or so black Angus, including what must be twenty calves. They are moving in the same direction as we are, along the bank, also following the river. Further inland – perhaps half-a-kilometre away – is another, higher, bank stretching roughly parallel to the one we're on, evidence of an earlier contour of this great meandering river.

Our path is cut by gullies. We scramble across them. The cattle pad disappears. We find another. A skeleton tree full of sulphur-crested cockatoos explodes at our approach. We turn our necks and watch the tree fill again as we pass.

The river is at our left, a marsh to our right. We see the herd of cattle ahead of us once more. The sight of them moving in the same direction is reassuring. River and humans and egrets and cattle all part of the same flow. This is the trick to life: not to think and plan and guess and second-guess, but to give oneself to the natural way of things, not to fight with the universe, but to submit to it. We fall into a walking reverie high on this embankment.

But then the mob stops. Heads turn. The cows eye us directly, fifty of them, in unison. The calves swing their necks, this way and then the other, stirring. The mob changes direction and begins to track inland, before turning again, this time making their way towards us at a trot. We move left, as near to the river as we can, and up onto the embankment, to let the herd pass. The ground reverberates as the beasts approach. They're not spooked. They're determined, this mass of snorting heads and spraying shit and pounding hooves. As the herd reaches us, they thunder down the embankment to our right, skirting the inland lagoon. Very, very strange. The mob protecting its calves, I think, from some threat, invisible to us.

Until, as we continue on our way, we understand. The embankment ahead comes to a dead end. The river has cut the embankment and flooded the lagoon that bends around the raised land, isolating us. Our way is blocked, and there is nothing to be done but to turn, retrace our steps and follow the cattle. Who know better than us.

~

With the lagoon behind us, we crawl beneath a barbed wire fence and make our way cross-country through pasture up to an even higher embankment. We follow the ancient contour of the river. Though, of course, it is now, in these flooded conditions, once again the true bound of the river. Sometimes it's only at extremes that we see how the past is the present.

The grass is high, and we are exposed. There is a farmhouse in the distance to our right, sheds, a small dam, a sag of telephone wires limping along the horizon. Ahead, somewhere, is Black Snake Creek. We angle towards a sunken, jagged tree line to the south.

The final approach is by a complex of gullies. After Ferny Gully yesterday, we're ready for this. But unlike yesterday, the country either side of Black Snake Creek isn't impenetrable forest; it is undulating and open with patches of sclerophyll eucalypt forest within the thickly grassed paddocks. We withdraw from the bank of the creek to the high paddock and make our way inland, where we track the creek upstream from above, scouting for possible crossings. The creek is backed up a long way. The first causeway is a couple of metres under. We round a rocky bluff. On the other side the creek narrows. Still too deep to wade across, but perhaps it's narrow enough to leap. As a schoolboy, Grantley won long-jump ribbons, so I take my cue from him. He shakes his head. We push on. Black Snake Creek widens again, fills its own tributary gullies, and too late, we find ourselves in a marsh. We squelch our way through, shin-high. Another prospect – an overhanging fig branch. But the bough is too thin, and climbing the tree with our heavy packs would be tricky. We press on till we reach a four-wheel drive track leading down to the creek and rising up the other side. While it's not possible to gauge its depth just from looking, the creek is wide, its surface still, and the sun is breaking through the early-afternoon cloud.

I enter first, descending into the water – ankles, shins, knees, thighs – feeling the bottom with my walking sticks. It's flat and solid underfoot. I pause midway, turn and give Grantley a thumbs up. It's good to be here. Precisely here. Standing in the middle of Black Snake Creek a kilometre-and-a-half upstream from where it joins the river. A hawk hovers below a cloud. The rippling creek, the fluttering wings, my pulsing blood. The water is cool against my thighs. I look down. Half of me has disappeared beneath the surface of the brown creek. But the other half looks back up at me. I hear the voice of Hermann Hesse's ferryman in *Siddhartha*. 'Have you also learned that secret from the river,' the ferryman asks, as if posing the question to me now, 'that there is no such thing as time? That the river is everywhere at the same time.'

In no time we are climbing the southern bank of the creek and sloshing through a field of high grass, negotiating barbed wire fences, following a dirt road and descending once more to the river. It is amazing that the sun is beginning to set and that there is beauty such as this in the world.

~

The flood has swallowed the timber jetty that guests at Josh's campsite use to launch canoes, but the river terrace itself is still a couple of metres above water level and there's no chance of it rising that much overnight. I check the capacity of the dam upriver: it's 107.9 per cent after four days of controlled releases intended to 'drain down the flood storage compartment, with ongoing releases likely to occur until Sunday'. Today being – it takes a while to place the day of the week – Wednesday.

Grantley goes down to the water's edge while I pitch my tent. He checks his messages while waiting for his lift back to Brisbane. His daughter is worried about a university exam, he has a business meeting early tomorrow, and a mutual friend is keen to see him for coffee before he flies back to Sydney. I turn my boots upside down on a rock to drain and hang my gaiters and soaked clothes from the branches of the lone gum growing in the centre of the campsite.

The sky is still infused with faint pink as Grantley's lift arrives and he takes his leave. He'd lingered as long as he could, precious minutes. The river rushes past in front of me. Grantley's there. And Ian, and Dominic and Steve. Crickets explode behind me. Stars begin to pop. The Southern Cross and its pointers are the first to appear. Those two pointers align perfectly with the direction of the river as it flows towards the city, showing the way I must travel tomorrow.

~

I wake in turbulent moonlight. The river may have outlasted the cicadas, but the moon has something to say. I crawl out of my tent. The eucalypts on the bank opposite are perfectly silhouetted, the moonbeams picking out each leaf, the river's diadem. I go down to the water and sit on a rock and dangle my feet over the tumult just below, and listen.

Day 19

Fairney View to Sapling Pocket Reserve

Distance: 19.55 km

Evening Camp: Place of the impenetrable forest

The first hours of day are filled with the mundane and the necessary. Is this what the moon was illuminating? The angle at which I'd set my upturned boots last night hadn't been sufficient to drain the water, so I reposition them. I shift my shirt and trousers and socks onto a different branch so they catch some rays of direct sunlight. The tent and fly could do with drying out too, so I hitch a line with spare rope from the tree to a boulder and drape them across while I wait for my journalist friend, John, to arrive. We'd first met on the sidelines of football fields, our conversations mediated by the ebb and flow of children's soccer games. Now we're meeting beside a flooded river, and who knows what it might mediate?

I scoop muesli into my bowl, sprinkle powdered milk, add water, stir them together and can't imagine a more delicious breakfast. Is it a failure in me that I am so easily satisfied? I boil water for coffee and return to my rock by the river, and if I'm waiting for something, it doesn't feel like it.

~

When John arrives, I stand with him on the terrace where I'd slept and dreamt, and together we look downriver. The bank is steep and thick with scrubby vegetation for the next kilometre. John's wife, Donna, has dropped him off. She looks downriver too, her face incredulous. She laughs, shakes her head and leaves her husband to his latest folly. I assess our options. Assuming we can get through, the route up the river will land us, eventually, at Sandy Creek.

On the satellite map, Sandy Creek is innocuous, but by now I know better. I ring ahead and speak with another John – John C. – once a lawyer, now a horticulturalist and a keen-eyed observer of the river in flood.

'Do you reckon we can get across?'

'Ah, no.'

The 'ah' was fellowship. As a young man, he too might have been eager to explore the boundaries of the impossible. The 'no' was unequivocal. The river is up nine metres in front of him as we speak, which means Sandy Creek will be backed up. When Sandy Creek is backed up, it backs up all the way to Pine Mountain Road.

So what to do? Even if we managed to make it downriver to its mouth, we'd have to begin a kilometres-long trek up the bank of the creek through an unknown – but likely flooded – landscape to Pine Mountain Road, where, after crossing, we'd need to make it back down the other side of the creek. It would be a painstaking introduction to the riverwalk for John. We'd never make our planned destination tonight. And, bruised by Ferny Gully and Black Snake Creek, I'm not enthusiastic. With my ankle swelling, I'm also tired.

'After weighing it all up,' I say to John, 'I think it's probably best if we just walk along the road to where it crosses Sandy Creek. It's only a couple of kilometres, and we can assess where to from there.'

John is a former China correspondent for the ABC, fit, sharp, worldly. I imagine him ruthlessly sacking fixers who hesitated in getting him into the valleys of northwest Pakistan after the 2005 earthquake. I imagine him thinking to himself now that my caution is pathetic.

'It's your show, Simon,' he says immediately, upbeat. 'I'm just here to support you.'

So he does think I'm pathetic.

~

We leave the river and follow the bitumen, deserted backroads mainly. A recently struck hare lies broken on the verge, on its right side, hindlegs outstretched, its left foreleg folded under its body, the fur of its belly pure white, its flanks and shoulders russet. Our nonchalant, guiltless killing. We pass the Pine Mountain Saw Mill. A kestrel observes us from a skeleton gum. Burtons Bridge over the river is cut. There are *Water Over Road* signs, black lettering on high-vis yellow. A corrugated iron shed, rusting blood red. The slate-grey sky.

We reach the bridge at Sandy Creek. Fist-sized white rocks are exposed in its golden, eroded banks. The water is very high, though we can tell from the debris-line it has been higher. We cross the bridge and follow the road through a patchwork of farms, angling our way by a series of doglegs towards the river.

A hawk hangs over every paddock, the head of each still and fixed as the pole star. The hawks have divided the fields among themselves and wait. After the weeks of deluge, these are hunting days, and John and I are no threat as we lumber predictably along the bitumen. In the throat of the breeze, a hawk drops out of the sky, gathering speed, pure intent. Hunter disappears into crop. Time and blood stop. Behind that screen of grass there is killing to be done.

~

These are small landholdings, each farmhouse nurturing its own dream. The newest and best of tractors are lined up neatly outside one. Snarling, half-fed guard dogs patrol the bounds of another. The barbed wire front fence of a third is garlanded with signs, *Corridor Conservation Agreement* and *Waterways Conservation Partnership*. A wooden chapel has been relocated so it stands shoulder to shoulder with a cottage, both with new green roofs, matching mustard-coloured walls, and lavender fringing the circular driveway.

'How are you recording all this?' John asks, by which he means the riverwalk.

I give him a list: journal, photos, audio recordings, the gathering of mementos. He nods: yep, the tools of his own trade.

'And what's your memory like?' he presses.

How do you answer that?

'Reasonable,' I say, and ask him about his. 'Reasonable,' he says, and we smile and talk, walking side by side along the road, about memory's imperfections and about ways of training it and about facts and about truth and about experience and how sometimes our bodies remember differently from our heads.

'But ultimately what I think,' I say, 'is that whatever we remember of an experience like this riverwalk is what we *need* to remember.'

'Love it!' John exclaims. And I think he might know what I'm trying to get at.

We come to a small park, once the site of a permanent water source, its name now marked on a sign: *The Bog Hole*. You don't forget a name like that.

Then, soon enough, we reach the river.

~

Sapling Pocket is the site of yet another quarry, though this one is abandoned.

Slowly I'm beginning to understand that the river isn't just made of water. It's made of sand and gravel too. I'm also coming to understand there's a breed of humans who look at the Seine or the Ganges or the Mississippi or the Brisbane and see not water but sand and gravel. They see opportunities to mine the beds and banks of their alluvial deposits – the stuff of concrete, which, along with steel – has fashioned modernity. Has built the city ahead. An entirely different history begins to unfold before my blinkered eyes: one of dredging and extraction and processing of alluvial materials. Of pits and processing plants. Of gravel-crushing and sorting and batching. Of branches of science dedicated to classifying the properties of various sands and gravels. (When *does* a piece of gravel become a grain of sand? Answer: When it is smaller than 4.75 millimetres in diameter.) It is a history of logistics: of supply chains and the costs of transporting materials to construction sites.

Concrete: that completely unremarkable mix of cement, sand, gravel and water. Unremarkable though revolutionary. And now, ubiquitous. After water, it's the most used substance on earth. We build cities with it. Economies are sustained by it. Development depends on it. Growth. Prosperity. Happiness. Human progress as measured by the volume of concrete pouring its way over our lives. We keep paving paradise to put up parking lots, as the song goes. Hard surfaces that lead to surface run-off, erosion and flooding. To urban heat islands. I want to rage, rage, against the loss of habitat and to take that rage out on concrete.

But don't forget the Pantheon, that early concrete architectural marvel. And the footings of the Eiffel Tower. And the guttering in the street that keeps rainwater out of our homes. And skate parks. And sporting arenas. Bridges. Libraries. The art galleries and theatres lining the south bank of my downriver city.

So if concrete is indispensable, we need to get our sand and gravel from somewhere. But not just any sand. All the world over, it is river sand that is king. All other varieties of sand bow to it. Marine sand is salty and requires washing before use. Desert sand is so wind-buffed that its grains are too round, too uniform – it doesn't bind like alluvial sand, which is coarser, sharper, more angular.

~

The Brisbane River has been plenty mined, but there's another reason alluvial material may be dredged from a river: to deepen and widen the channel. Usually both goals are pursued in the same operation. The Brisbane has been dredged from the first days of white settlement. The river was the colony's principal transport route. It needed to be 'improved' for the steamers, sailing boats, barges and market boats that plied their trade upriver, as far as the junction with the Bremer and up it to Ipswich. It was too shallow in parts, too winding in others. There were sandbars to dredge, rocks to detonate, bends to straighten, retaining walls to erect. It was the duty of 'civilisation' to canalise the river. To make it navigationally more certain. More efficient. The river needed to be trained. Tamed.

Downriver from here – in the reaches from Mount Crosby through to the city and out to the bay – the river is, in parts, unrecognisable from the one that flowed through that country before colonisation. The sandbanks between Petrie Bight and Kangaroo Point that at low tide could almost be waded across, gone. The Eagle Farm flats, dredged. The mudflats at Hamilton, filled in.

The river was dredged to deepen shipping channels, re-dredged when the channels were filled by flood-borne silt, and then dredged nearly continuously after that for a hundred years and more. The scale of sand and gravel extraction is sobering, the numbers difficult to countenance. Between 1878 and 1886, the *Groper* took out two million cubic metres, and between 1882 and 1886, the *Octopus* scraped out nearly three million cubic metres. Still the river had more to give. A century more, until dredging was halted in 1996, after more than 173 million cubic metres of material had been taken. The environmental consequences of a deeper, wider river?

Bank erosion, loss of pools and riffles, greater turbidity, greater tidal range, increased tide-speed and salty water reaching further upriver, to say nothing of the biological effects.

Even now the river mouth continues to be dredged for the port. Maintenance dredging, carried out by the *Brisbane.*

~

John and I stop for lunch on a ridge at the northern entrance to Sapling Pocket Reserve. I hang my sweat-soaked shirt over a branch to dry in what might be a breeze. Ahh. Tired legs. Sun on my back. Midday wrens flitting around us. Anthills. A beehive. We offer each other food, count the kilometres we've come so far today, and talk about where we want to get to tonight, a cabin owned by a riverwoman, Anne, at the evocatively named World's End Pocket.

'How do you know Anne?' John asks.

I'd met her in the early days of planning the walk. Anne is in her late fifties, diminutive yet powerful, friendly, open, and she was quick to offer me the cabin on her horse property. It'll take a solid afternoon of walking to get through the reserve to her place, so we need to get going.

We stride up an unsealed private road leading towards the river. Not so long ago it carried convoys of rumbling trucks filled with sand extracted from the riverside pits up ahead, taking their alluvial treasure out into the world and its markets. Although those trucks are now garaged, they've left their mark: parallel tracks of compressed earth snake through the high-grown grass. The road hugs the hill, rising gradually. We slow. Ahead the road has caved in, most of it fallen away into a steep gully, our further passage prohibited – or not recommended. It's hard to make out exactly what the sign means. *Unstable Surfaces.* The recent erosion is fresh, so we tread carefully along the narrow ledge that is all that's left of the mining road.

On the other side of the collapse, the road descends. *Slippery When Wet*, below a pictogram of a teetering truck within a red triangle. But the view to our left across a canopy of remnant rainforest reveals a straight reach of the river, coursing directly towards us. Though the river is full and fast and wide, we are high and untouchable. The further we descend, the more the forest begins to stretch over the road, reclaiming space that was once its own. At the tip of the peninsula the river curves expansively to the right, swamping the low inside bend and submerging the remnants of the old gravel pits. Relics of an industrial Atlantis. I toss a stone into the swirling river. Splash, gone.

The clouds are breaking up and beginning to skip across the blue. In the river, it's as if a forest of trees is growing from the caramel-coloured water itself, shuddering in the current, frightened. Three bright blue plastic forty-four-gallon drums have become lashed to one of the trees. They pull desperately, frantically attempting to liberate themselves from their short leashes. Perhaps they sense their fate. When the water level drops, they will hang from their scaffolds, limp, defeated.

Some way downstream, we come upon a natural debris trap higher on the bank, souvenirs from February. Steaming mounds of dried grass, current-twisted. A wooden gate, VJs, an upturned bentwood chair, a round hay bale, muscular logs, denuded branches, the usual plastics – capped bottles and pallet strapping and shopping bags. I stand before the debris for long minutes. There's a pattern to the way these objects have been strewn by February. The way they've been arranged to rest together, each beside each. Some language of the discarded I strain to hear.

~

We pick our way quickly through the debris trap, each misstep sinking us into the thick jetsam whose bottom we can't measure.

Both of us are thinking, but neither is prepared to say, that this great island of refuse has probably become a giant snakes' nest.

The puppeteer sun works with the northerly breeze to pull the clouds higher. We stride out. The way around the tip of the reserve is surprisingly easy, despite our intended route being submerged by the floodwaters. The new route we've found – just above the current water level, but below February's floodmark – is perfect. February had drowned a wide band of plants and grasses. Now, that ribbon of dead grass and low castor-oil plants makes it easy going. Long strides, wide flowing river to our left, forested nature reserve to our right, breeze gently blowing into our faces. We are our surroundings. My spirit is buoyant.

But this is a long reach.

And it's wrong to think that a forest reserve left alone for years to regenerate is necessarily a riverwalker's friend.

~

The forest creeps closer to the bank, crowding us out. If it wasn't flooding, there would still be plenty of bank to walk, but not now. We stay low at first, climbing over branches, squeezing between trunks, grasping liana vines when we lose our footing. Then we come to the first creek. I know the drill and am already quickly scanning the terrain and assessing whether we can cross here or need to head further upstream for a passable route. But John needs to process this. The descent to the creek is steep and slippery, degrees of difficulty beyond the assistance our walking sticks can practically offer. But the forest is thick with trees to grasp or to rest our shoulders against while we catch breath. We step through the dark water at the bottom of the gully and then pull ourselves up the opposite bank, trunk by trunk. We pause at the top, panting, sweating. John is grinning. He absolutely loves this.

We press on.

The forest thickens, thins, thickens, thins again, a pulse of sorts. I settle into a rhythm, beyond thought, nearer submission. There's another creek at the bottom of another overgrown gully, and a third after that, and a fourth. Some are crossable near the river, others force slow-scrambling expeditions inland.

Sometimes we leave the forest and drop to the water's edge. When we speak, it's about the hope of finding a track down there. What we don't say is that the forest is suffocating, and we need to escape it to breathe. But there's never a path by the water, only mud, and the relentlessly racing river. We wade through the mire for a few metres. It's slower going than through the scrub, and even more exhausting. So we return to the forest.

Clouds overwhelm the sun by midafternoon. I check my watch. I count the creeks we've crossed, estimate the distance we've come, calculate our rate of progress and guess how far we've yet to travel to reach Anne's cabin. It's touch and go whether we'll make it before dark. But the country should open up as we get to the end of the reserve, meaning the going will be easier. We've little choice but to press on: in the last couple of hours we haven't seen a single clearing large enough to pitch two tents.

The scrub thickens even further. I stop at a wall of vegetation. John is some distance behind me, and slowing. He's beginning to labour. I wait. He's breathing hard, but I don't want to look him in the eye. Though the forest seems impenetrable, it's not – there are tracks through, but they're low, small animal paths, wallabies or pigs or bandicoots. I show John how it's done, on hands and knees. Because there's no other option, no going back. And because years of crawling along bandicoot tunnels through lantana as a kid taught me that a path is a path.

We move through the forest like this, crawling and scrambling and sliding. One moment we're disentangling our backpacks from a lasso of lawyer vine, the next we're clutching at a thick loop of liana when we lose our footing. What is friend and what is foe depends on whether or not it aids our passage.

'I need to rest, mate,' John says when he reaches the strangler fig where I've been waiting for him.

'Of course, mate.'

The 'mate' hides, for both of us, the gravity of it all. As fit as John is, moving through terrain like this is different. I've been walking and scrambling with a twenty-kilogram pack for a few weeks now, and my body is hardened to it. John collapses onto a low buttress root. Sweat pours off his glistening forearms in sheets. He stares into the mid-distance, his eyes wide, bulging.

'Simon, I don't think I can …'

I cut him off. I'm not yet ready for defeat.

'What we'll do is this: you give me your heavier gear. I'm still feeling strong. My body's had practice, and my pack has plenty of room in it. Shed some weight, John, and it'll make a huge difference. Just watch.'

I've been unshouldering my backpack as I speak. I open its mouth wide. 'All right. What can you give me?'

John gets out of his pack, and unclips it. There's an Aladdin's cave in there. First he produces a large transistor radio, 'to listen to the election coverage on Saturday night'. What journo isn't a political junkie? Then a bottle of Canadian Club. Also for the election. A bottle of red wine. 'A gift for Anne.' The usual mix of camping food and cooking gear. John sheds about seven kilograms. I load all of it into my pack, all except the red wine, which simply won't fit. I rest it on a patch of grass near the buttress root. He's a good man, thinking to hike a bottle of red in for Anne.

'She won't expect a gift, John. Seriously.'

He's beyond making decisions. He just nods. We clip up our packs, shoulder them and both stand. John nods again.

'Yep, that makes a difference.'

It's 3.30 pm. It'll be dark at 5.00 pm. We resume our journey.

It's not so much the extra weight that worries me. It's that my pack is sitting higher now, well above my head. My centre of gravity has changed, and everything feels off. I lead the way down a gully, a litter of red and yellow leaves on the muddy surface, momentarily too pretty to step upon. I have to duck lower now to get beneath branches, but my muscles aren't tooled for the changes my body now has to make to accommodate my pack's new shape.

We come out of the forest and follow an animal track through a patch of grass on the sloping bank. A miscellany of trees: leopards and Moreton Bay chestnuts and small-leafed figs and bluegum saplings and *tessellaris*. In the distance is a crow's apple, *Owenia venosa*. I think. Or maybe not. The forest and river squeeze out the grass. I force myself into the dark tangle once again, pulling John along. Hope and folly tease each other. Step follows step. I stumble under my new load. John's second wind is blowing out, and the afternoon is as good as done. I lift my water bladder to my lips, but there's only the suck of air. I'm out of drinking water. The low cloud can hang on no longer and splits open, pouring itself out onto the forest canopy above. We're done for the day, all done.

~

What we settle on is not a clearing. Nor is it flat. It's a ribbon of sodden bank falling away towards the river. The ground is too boggy for tent pegs to get any purchase, but at least there's room for two tents. The river rushes by, closer than ever before. The rain spits on our backs as we huddle over a fire and warm ourselves with hot meals. Hot and gritty – we've rehydrated our vacuum-dried packet food with floodwater, thick with sediment washed down from the Lockyer Valley.

I fill my bladder with the brown river water too and drop in purification tablets, then slake my thirst with a water-and-dirt broth.

I'm too thirsty to wonder about the soil washed down from those Lockyer Creek farms. Too thirsty to worry about the nitrogen and phosphorous in it.

We retire to our tents, and I send Anne a message by satellite that we've stopped for the night in the reserve and will reach her tomorrow.

'Stay safe,' she messages back.

My tent is pitched so it's aligned with the river, head pointing upriver, feet down. I lie on my air mattress, and immediately my body begins to slide down the slope. It's impossible to sleep like this. In an attempt to create a flatter surface, I jam my rolled sleeping bag under the downhill side of the air mattress at torso level, propping it up. It works.

I'm sweaty, dirty, muddy. My clothes are wrapped in a plastic bag outside. I crawl into my sleeping sheet, lay my head on the puffer jacket as I have every night, and fall asleep in a crudely cradling mattress.

I discover during the night that the slope doesn't simply fall from right to left, but also from head to toe. With my weight upon it, the mattress slides down the floor of the tent. My feet press against the wall, growing damp. I wake. The temperature has dropped, my feet are wet, and I've grown cold without a sleeping bag. I put on my puffer jacket and shove all my clean clothes into the jacket's bag as a new pillow. I'm thirsty and have to drink. While the particulate in my water bladder has settled a little, it's still crunchy.

I wake once again during the night to attend to nature's call. The slope is slippery as hell, dangerous. I steady myself and face the current. In front of me, across the river, stands a single ghost tree, its limbs reflecting moonlight like bone.

Day 20

Sapling Pocket Reserve to World's End Pocket

Distance: 5 km

Evening Camp: Place where the world ends

Something has changed, some shift, as if the world is, this morning, different from the one I'd closed my eyes on last night. I can't fathom, at first, what it is. Perhaps it's just a dream, heavy still on my chest. I clamber out of my listing tent.

John has the fire going. He's wrapped himself around a hot mug. 'Morning.'

And then I see: the river has dropped a full metre overnight. The ribbon of mud and drowned grass is a metre wider this morning. It seems impossible the night can have such power. But it's not the moon or the skeleton gum across the river or the god of clouds that has pulled the river lower. Late last night, the dam operators cut the outflow from the dam, to allow a more gradual lowering of the water level, reducing the risk of the riverbank slumping. Amazing. Flick a switch, stop a river from flowing. Just like that.

~

There's no mobile phone coverage down here, low on the bank, along a stretch of river without a building in sight, a ridge behind us. But orbiting the earth well beyond the clouds are the satellites that have taken pictures of the entire planet, including this reserve, and that connect us near-instantaneously to databases of those images. After yesterday's slog, we want to see those images, to know how much more of the same we have ahead of us. We want to know what's on the other side of the reserve, and how we'll walk the last leg to the end of the peninsula and World's End Pocket. We call up an image of this length of river. When it was taken I can't tell. The river in the image is slim, unperturbed, still. If I had access to other satellites recording the surface of the earth this very moment and transmitting their spying in real time, I might recognise the river as swollen, might see two unknowing figures beside their two forlorn tents. We must satisfy ourselves with this dated image – though even back then, in those days before the deluge, this pocket of the planet was dark green country, darker than any vegetation for miles around.

What the map also shows, ahead of us downriver, are two thin lines threading through the jungle. They look overgrown, but those lines might yet be tracks we could follow.

~

I climb into yesterday's clothes, lace my boots, clip on the gaiters and pack away my heavy wet tent, the sleeping bag and mattress and sleeping sheet, all damp. I look back at the rectangular imprint I've left upon the earth. We move slowly along the bank, making use of the additional clear ground that the dam operators have provided, tracking the ribbon of freshly drowned grass, reluctant to go back into the forest.

The contour of the jungle's canopy changes. A good sign. We climb the bank to a flat river terrace. It's high-grassed, but treeless. This is ground that was once heavily worked, whether by industry or agriculture I'm in no mood to discover. Not now. Now, I only have antennae for the track. Any trace of it will do, the merest hint of a way, of human passage through this overgrowth. And sure enough, there it is, exactly where the satellite map promised, the overgrown but unmistakeable parallel wheel-ways of vehicles. Comforting way-lines running riverways along the terrace.

We take the track and head east between the flood-loud river and the jungled hills of what's called Cameron's Scrub. The track forks. We take the river-most fork. It drops off the terrace and leads unhesitatingly towards the water, descending steadily into the river itself as if it has always done. That way leads towards tests of faith, whole-body river immersions, baptisms into bold beliefs. We backtrack. The right fork climbs steeply for two hundred metres. Low cloud. High humidity. Volumes of sweat. The steep hill silences us. We don't turn until we crest it. When we do, hearts pounding, the only thing we can see is river – left and right and ahead. But we can no longer hear it. And we are thirsty. I twist the valve on my bladder, close my eyes and taste the river's grit once again.

In time we follow the track down the other side of the hill, and the river disappears once more. The walking is hard – the track is steep and the air sticky. But we want to walk. We want to leave the suffocating forest and yesterday's ordeal behind.

We check the satellite map again. This ability to see precisely, and so easily, where we are, to have this knowledge at our fingertips, is transformative. John and I are walking along an overgrown four-wheel drive track in an environmental reserve unknown to all but a few, visited by mere handfuls, and yet here it is on a GPS image. Here we are on that image, moving slowly but unerringly across the screen. The same icon pins us to the earth whether we're walking here or leaning from the Pont Neuf over the Seine or fishing for eel on the Whanganui or diving for sand from the bed of the Niger. The democratisation of this capacity to locate ourselves is giddying.

But what do we lose? Our very survival once depended on knowing where we were. We read the stars, or the winds, or the land, or the migrations of birds and butterflies, and knew where we were. Just a decade ago, to visit a new location we'd have to look at a map and then memorise the route – left here, right there, straight ahead across the river – and then follow it, all the while intently alert to the streetscape. Trip by trip, our knowledge of our world expanded; trip by trip, we learned the geography of our lives. Now, I enter an address and am guided by a GPS-armed voice to my destination without ever needing to know where I am, and I worry: am I losing my ability to find my way in the world? To know where I am? And, if place shapes identity, who I am?

~

Ahead of us, a busted twenty-thousand-litre concrete water tank stands like an outpost of a collapsed civilisation, at the head of a driveway turning off the forestry track. A fig tree overwhelmed the tank a long time ago. Its limbs reach out through the holes it has punched through the tank's thin walls, revealing its ineffectual structure: two-and-a-half centimetres of poured concrete, a cage of reo, a layer of tar. Now cracked, stained, rusted, mossed, lichened. Down the driveway is an abandoned house. We leave it to its fate and continue along the track.

Around a bend is a small clearing and three gleaming white four-wheel drives, silent. I stop, point my walking stick, turn to John and raise a finger to my lips. He nods. We approach slowly, but there is no-one with the vehicles. The decal on their doors and the tanks and wands in the trays tell us that whoever they are, these people are not tending a hidden marijuana crop somewhere in the forest, but are here to spray weeds.

Once upon a time, not so very long ago at all, country like this was maintained quite differently. There was no need to spray to keep a forest under control. Back then, country wasn't locked away in reserves. Back then, the bush was fired. Regularly, carefully, knowingly. Fire knowledge was applied to land to maintain the health of animals and plants, to reduce wildfire risk, and for hunting and ceremony. A question, one to join those that Steve asked on the first day of this journey when wondering what the country used to look like and what it used to *be* like: might practices such as Indigenous burning help do the work of healing unbalanced landscapes?

~

We emerge from the forest onto Riverside Drive. To the left the road leads to the end of the promontory, to World's End Pocket, and to Anne's cabin. To the right it follows a ridge running parallel to the river. It is unsealed and regularly cut by rain. In parts it's cloistered by forest, in parts kept company by riverside farms. We know we're not too far from the cabin now, an hour, no more than two. We stride it out. It rains lightly for a few minutes at the top of a rise, the road overhung by strangler figs and crows' nest ferns. The forest opens onto a valley to the left, a swell of vegetation rolling down to the river in the mid-distance. But there is nothing Arcadian about this view. Rather, the plants in the valley are covered by vine, as if a giant net has been thrown over every hillside and every gully, a web of dark green smothering the land, only the tallest eucalypts and hoop pines capable of piercing this asphyxiating shroud.

It's not one of the usual culprits – cat's claw or potato vine or madeira vine. The leaves of this one are broad, in leaflets of three, some heart-shaped. The stems of the vine are narrow, tough, twining. Listen carefully and you can almost hear it grow.

At a saddle in the road, we come upon an old man on a ride-on mower, stationary on the sloping verge. He's the first person we've seen today. We wave as we approach. He returns the greeting and smiles. He's in his late eighties, in grey shorts and a work-stained blue-and-red checked short-sleeve shirt. His reading glasses are tucked into his breast pocket.

'I'm stuck,' he says airily.

Two of the mower's wheels are hanging over the verge and can't get traction. He figured it'd be risky to try to dismount – the mower might tip, and he'd hate for it to roll on top of him. He's hosting a family wedding tomorrow and has tried ringing the groom, but the young man is in the final stages of planning and isn't picking up.

John and I push the man out of danger, and we get talking.

'You've got to speak with my neighbour, Charlie,' Anne had said on an earlier visit to World's End Pocket. 'There's nothing he doesn't know. I'll ring him when you arrive, and introduce you to him.'

Here Charlie is now, mowing the verge for his wedding guests. We settle in for a yarn by the roadside. He tells us who the early property owners were, what land changed hands, when and to whom. The groom arrives in his four-wheel drive, but Charlie waves him courteously away – whatever emergency there might have been has passed, and there is conversation to be had.

I ask about the vine.

'The glycine?'

'The one covering the hill back there.'

He nods, glycine.

'Cattle feed,' Charlie says. 'Back in the day, they said it was fire-resistant, drought-resistant. But you need a lot of cattle to keep it under control, and these days ...'

I don't know what to say. So this is what the road to hell looks like. The suffocating effect of good intentions, the irreversibility of folly.

'Noogoora burr is the worst, though,' Charlie says.

My interest is piqued further. In my pack I've got the specimen I'd collected with Ian from the upper reaches of the river. No doubt the burr is brutal.

'I've had Noogoora burr so high, I couldn't look over the top of it if I was on a horse. Nothing grows. You've got no grass, nothing. Noogoora burr overruns a place.'

Noogoora burr. I wonder what Charlie might know about the history of it.

'They say the burr was named after a station somewhere on the river—' I start, but there's no need to finish.

'Noogoora Station is just upriver of Sandy Creek.'

The same Sandy Creek we'd crossed yesterday. That sense of time. The way history endures through memory. Or as Faulkner put it: The past is never dead. It's not even past. If once there was a station upriver of Sandy Creek, there will always be one there.

'I was told when I first came here,' Charlie says, 'that you could count on three floods every year, and that will drown the burr. But that was before Wivenhoe.'

So we talk, inevitably, about flooding: 'All those millionaires down at Graceville worrying about their mansions.' And quarrying gravel: 'When the pirates left the sea, they became gravel merchants.' And the hoop pines on the top of Pine Mountain: 'There are still some good sticks in there.'

When it's time for us to continue on our way, I ask whether he minds us walking along the river in front of his place if we need to.

'No,' he answers immediately. 'I don't mind if you walk through. I shared the river with people over the years, and they shared it with me.'

~

We arrive at Anne's place at midday, bedraggled, tired, sore. We head for the cabin she's offered us, on the other side of the property from the main farm house. We strip to our jocks and hose ourselves down outside the cabin. We wring the rain and sweat and tank water from our clothes before hanging them over the fence. Prop our boots under the tank stand. Step carefully into her immaculately decorated cottage, wiping grass and mud from the soles of our feet. Waiting for us inside is leek soup with melted stilton cheese, and baguettes on the side. It is a lunch for princes being offered to river hobos with nothing – after I'd forced John to abandon his bottle of red – to offer her in return. Does she not realise how dirty we are, how undeserving? We'll join Anne for dinner up at the house this evening.

After we eat, I lie down and sleep. It's more than a nap. I disappear for an hour or more, and cannot fathom, when I wake, what lands I visited while I was away.

~

It's been raining for weeks. It's been raining here on earth for 4.5 billion years. Whether water was here from the start, formed as the primordial planet started to cool, or arrived on the backs of comets or proto-planets from the outer asteroid belt, rain has been falling on the surface of our planet and pooling and trickling and flowing and leaving its mark for a very long time, sculpting the earth as it goes.

And it has been gathering, molecules attracted to molecules, for a long time too. In creeks and rivers and lakes and oceans. Frozen in glaciers and ice caps, as if resting. Seeping through soil and rock to pool in subterranean aquifers. Flowing, always, with gravity.

Until, that is, it rises. Until it evaporates from the oceans and seas and lakes and dams and rivers and passes overhead in cloud.

And then returns to rain again.

Everyone has been following the rain. I sit on the front steps of Anne's cabin and respond to text messages. Alisa. She'd love this cabin, this property, this cradling river. I send a note to Steve and Dominic and Ian and Grantley, letting them know where we are. Another note to the group of canoeists who are keen to paddle with me around the prisons at Wacol, and to my boatie friend with the Moreton Bay cruiser who'll be my companion for the final river leg, down to the mouth and into the bay beyond. The truth is, I say, who knows what the river will allow and what it will forbid?

I sit on the steps of Anne's cabin and watch the river flow across the wide, wet land. It's been flooding so long now it's hard to imagine the river is capable of different moods, has ever had a different life. The intensity of the last couple of weeks of rain – the immersion of the present – is slowly eroding a lifetime of river memories. I close my eyes and try to recall what it looked like here six months earlier when Anne led me down to its banks. I remember: there was an island in the river, a place of picnics and weddings and grazing cattle and flashes of azure kingfishers. There was the lungfish pool. And a new *Vallisneria* bed, where researchers from one of the city's universities had planted a patch of the river grass. 'Val' stabilises the bed, absorbs nutrients and offers habitat, and Anne offers the river readily to researchers and water-quality measurers and writers and river-retreaters. I hear her neighbour Charlie: *I shared the river with people over the years, and they shared it with me.* Generosity's infectiousness.

~

On the kitchen wall of Anne's house are mounted photos. Children, beloved racehorses, a river in spate – the things in your life you'd frame. There is an aerial photo of the house in January 2011, surrounded by water. The island in the river, Anne tells us, has changed since I was here last year. Back then – she says it as if it were a different historical era – the main channel was to the left of the island, but February changed that. Now the main channel is on its right, which means the island is no longer accessible from her side of the river, as it has been ever since she's been here. The events you'd mark on the chronology of your life.

We move into the dining room and sit at the long polished French oak table. Anne pours shiraz; we toast John's humble bottle of red lying forlornly on the jungled bank in the nature reserve, and the talk drifts slowly away from the river.

Day 21

World's End Pocket to Mt Crosby Weir Nature Refuge

Distance: 17.97 km

Evening Camp: Place of the escarpment of solitude

The rain returns an hour and a half before dawn, steady on the cabin's tin roof. I reach for my phone, but there is nothing to be optimistic about in the day's weather forecast. Overnight, the Bureau of Meteorology has announced that this is already the wettest May on record. The bridges downriver are all closed – Kholo Bridge, Mount Crosby Weir, Colleges Crossing – and the cross-river ferry service at Moggill has been suspended. And still it rains. I lie awake, lost in the dark labyrinth of changing climate and shrinking habitat, and the prospect of another day's walking in the ever-wet.

When dawn breaks and the first kookaburra of the day laughs, it is a relief, pulling me out of a threatening despondency. Great may be the consequences of our folly, I think, but still there is a kookaburra to laugh its warning.

Anne is up and grooming her palomino while John and I brew coffee and map out our route for the day. It is merely a guide.

Conditions underfoot determine everything. I text ahead to a friend who lives a couple of properties downriver, Jos, who's offered us morning tea as we pass by:

We aim to leave here at 7/7.30 am. How long do you reckon it'll take to get to your place?

Jos soon replies:

At least 4 days after struggling thru the mud on the riverbank. But if you're walking down Riverside Drive maybe 1–2 hours depending if you stop to admire things.

~

There's much to admire as we farewell Anne and her farm and set out for the day. Anne's optimism. John's resilience. Jos's dry humour. And, as we lean into the rain, I'm impressed by the grip of my Scarpa boots on wet gravel and how waterproof my backpack cover is and the genius of the person who first mixed sultanas and nuts into a hiking snack. And, as we walk into the future, the care that river observers and knowledge-keepers bring to their work.

~

Two hours later, John and I step into Jos's stables and out of the rain.

We unbuckle our packs, look around and find ourselves suddenly immersed in a world of horses. Hay, chaff, leather, horsehair, manure, piss and snorting from flared nostrils. A dentist is working on the teeth of one of Jos's favourite mares, a carefully groomed chestnut with a white blaze and star-bright eyes. She's the last of the morning to be treated. The rain is drumming on the high iron roof like the hooves of galloping stallions, and we are momentarily transported.

Horses have been threaded through the life of the valley these last couple of hundred years. The horse Commandant Patrick Logan was riding when he fell on the banks of the creek we crossed days ago and which now bears his name. The horses still used to muster cattle in the hills in the upper reaches of the river. The horses Andrew at Fernvale rides in cattle-drafting competitions. Anne's racehorses.

Surrounded by horses, footsore, as the rain continues to fall, my mind wanders to stories of flooded creeks crossed on horseback. There's a mythology, almost a trope, of horses and flooded streams, the rider desperate to make it home to shelter from the storm. In those stories the creek is often one the rider knows well, has picnicked beside, drunk from, sketched, watered cattle at. It's a friend, and it's easy to think it's still a friend even when it's rising. And horses, riders believe, are made for crossing flooded creeks. A horse has instinct. It knows whether a creek is passable or not, six thousand and more years of accumulated creek crossings in its blood. From the steppes of Central Asia, where the first horse was domesticated, to the Snowy River, to Billy Mateer in this river valley in 1893.

But too often these stories of horses and flooded creeks end in tragedy. There are two in my family, and another in my wife's.

~

We adjourn to the house for coffee and raisin toast with Jos and Al, a wiry American former fighter pilot. Jos hands us towels to wrap round our shoulders to dry off. Her blue heeler begins to warm to us. The rain falls. We talk about rivers and danger. Al had been secretly ferried across a great South-East Asian river under cover of dark another lifetime ago. The American military weren't yet in Indochina. As local suspicions rose, Al needed to evacuate back across the river, which meant travelling by night and digging trenches to lie in by day, covering himself with foliage until he reached the river's bank, the longest seventy-two hours of his life. This morning we look out at a flooding river, wild, dangerous. Al asks whether we want another coffee. He returns to the subject of South-East Asia. Back then, he says, that other river meant safety. Al has a soft spot for rivers.

~

So rivers connect, but – like the river Al had been ferried across – they also divide. They're natural borders, keeping people and countries apart. The waters of some of our great rivers are witness to some of the most distressing transnational tensions of our time. The Rio Grande, between Mexico and the United States. The Jordan, between Israel and Jordan. Fluid as nation-states and national borders are, something like twenty-three per cent of the world's non-coastal national borders are delimited by rivers. And when there is competition for finite water resources, a river can become a source of conflict. The dispute between Ethiopia, Sudan and Egypt over the Nile; tension between Turkey, Syria and Iraq over the Euphrates–Tigris basin. Closer to home, there are the interminable disputes between Australian states over the water in the Murray–Darling Basin. Nearer still, I think of a story I heard a year or so ago about two landowners on this river: a downstream farmer accusing his upstream neighbour of drawing too much water during drought, angry words leading to baited dogs leading to intergenerational feud. True or not, I can't say, but the story seems to echo a depressing reality. That for some of us, a water conflict is worth waging war over.

~

'Follow the terrace,' Jos says. 'You can walk in front of my place all the way to Kholo Crossing.'

We make our way down from her home, through horse paddocks, down through a small patch of remnant forest that she protects fiercely, over barbed wire and then strands of electrified fence, to the river. I pull the hood of my rain jacket back momentarily. The river's surface is marbled by currents of silt forming patterns, breaking up, reforming anew. The high terrace is puddled, the grass horse-trimmed, the rocky outcrops sleek and darkened with wet. We set out downriver. I lead the way. It's good to be here, on our way once again, as if setting out is a natural state.

Across the river, houses perch haphazardly on the ridge, appearing and then disappearing behind veils of eucalypt and mist, their silver roofs flashing intermittently like warning beacons through breaks in the forest. The earth has slipped in front of one home that is now precariously positioned, leaving mounds of soil, uprooted trees and a naked fall. People have gathered on the bank before the home, an urgency of conferring engineers and landscapers and a cleave of neighbours praying, thinking *There but for the grace of god.*

~

By the water's edge, we spy a bright orange onion sack, twenty kilograms net. But it is empty and – washed downriver by the floodwaters – twisted and forlorn. Sewn onto its side is its badge of ownership, blue and red and yellow print on a white background: *Riverside Farming, Class 1 ONIONS … O'Reilly's Weir Road, Lowood, QLD.* It's like reading the nametag on the collar of a lost dog. But one I know. I'd walked past that farm, and those onion fields, five days ago with Grantley. I stash the sack into my backpack. When this is done, I might take it back to its Lockyer Valley home.

~

Nothing stays the same. The terrace we're on narrows, cut first by a shallow fold, then a deeper one, before disappearing entirely at the first gully. We track higher till we reach a boundary fence running along the top of the high bank, parallel to the river. On one side of the fence is scrub and chest-high grass and rough, slippery sloping ground; on the other is neat rolling pasture, beautifully parcelled for crop or stock rotation. We stay on our side of the wire in our rough, overgrown world. We trudge through thickening wildness, stumble over concealed logs in the long grass and wince as low-snapping branches draw blood from our necks. We try to ignore the tamed paddocks over the fence beside us.

As we approach the next gully, we hear the sound of an engine, a four-wheeler, approaching from somewhere behind us. I turn but can't see it. The rain and the fog and the wet hills distort the sound, but as best I can make out it's probably making its way towards us from the machinery shed we saw ten minutes back. We've been exposed, walking up high like this, and would be visible to any farmer who happened to glance up from their work and look across towards the river. In my mind I scroll through the list of property owners I'd spoken with along this reach of the river, and try to work out whose place we're walking in front of now. They'd all been enthusiastic about the riverwalk, had all offered a cuppa or a chat on our way past, and though I'd not notified every property owner of the precise date I'd be walking along the river in front of their place, I'd said May, and everyone was fine with that. Word might have already passed from Charlie or Anne or Jos – whoever it is making their way towards us is probably coming with a greeting. Or, if they're curious about who we might be, they'll no doubt remember when reminded of my earlier conversation with them.

So there's no logical reason to feel anxious, but I do. Something about the disembodied sound of the engine is foreboding. In this moment – overtired and immersed in rain and flood and overgrown riverbank vegetation – that approaching motor represents everything I'd sought to leave behind on this pilgrimage. In this moment, I desperately do not want to be found by civilisation or society. In this irrational moment, the motor becomes hunter and we its quarry.

Quick! We hurry forward and plunge down into the gully, where we are immediately out of sight. But still the sound of the engine comes. If we can just get to the other side of the gully, we'll be safe; the four-wheeler can't follow there. But the gully is very steep, its sides thick with lantana. We take an exposed route to the bottom. If the four-wheeler's rider reaches the top of the gully and looks down, they'll see us. But there's no way across the gully down here –

it's too wide, and the coursing water is deep and treacherous. We could press further up the gully to find somewhere to cross, but that would leave us even more exposed. The throbbing engine is nearly upon us. We scramble on our hands and knees behind a clump of lantana, and stop, panting. What we're doing – there's no denying it – is hiding. Like trespassers or thieves. The four-wheeler stops at the top of the gully. The pitch of the engine drops to an idle. A purr, patient. We wait. Our tracks through the bush above and the trail we left slip-sliding down the side of the gully must be obvious to see. We wait, hearts thumping.

Is that a voice? Do I hear someone calling out? Or are they talking to themselves? Cursing? Wondering aloud even, bemused that two hikers would descend into a gully like this, in weather like this? My heart races even faster. But whoever is up there does not follow. Does not care enough. They mount their four-wheeler and give it throttle and turn away. Only when it is quiet again, and the only sounds are water – creek and rain and leaf-drip – do we rise from our rude hiding place.

~

The route along the river from the other side of the gully is by the muddy bank. We pick our way slowly, driving our walking sticks deep into the mire, boot-deep step after boot-deep step, navigating trees and fallen logs and each succeeding gully. Vines wrap themselves around my backpack or ankles. I wrestle myself free and turn to warn John in case he hasn't seen them. The rain is very heavy. My glasses become so wet and fogged that I can't see properly. Drying them is no good – they get wet again immediately and my drying cloth is soon soaked anyway. But if I take my glasses off, my short-sightedness means I can't see far enough ahead to track the best route along the slippery bank. Either way, I'm half-blind.

John takes the lead, and I follow, placing my boots in his footsteps, not thinking further ahead than that. I'm tired. John is strong and decisive. He and I have swapped our roles of yesterday. I follow him and I'm grateful.

~

We follow an easement off the river and come to Riverside Drive once more. Rivulets pour off it. The road should take us most of the way to Kholo Crossing. We cross Coal Creek, another one, named in an era when the mining of coal was noble labour and creeks that hinted at deposits of the black stuff were thrilling discoveries. We pass the riverside pumping facility for a nearby coal-fed power station, the words *Clean Co* on a sign hanging from its locked gates. *Council employees and authorised persons only.* Though the power station is out of sight, thick transmission lines loop their way across the sky high above us, tower to tower, carrying power from the station out into the world. The towers offer a short cut. We veer off the road and walk beneath the humming transmission lines for a kilometre, currents of coursing electricity flowing above our heads.

Three Hereford bulls wait for us as we approach Kholo Road, grazing in the no-man's-land beneath the transmission lines. We stop when we see them. They're downhill from us, and raise their three great heads and eye us off. What effect have the rivers of electricity above them had on their mood? Does it agitate or numb? What effect does living in an electrified world have on any of us? We move forward carefully. The prospect of the bulls charging uphill at us is unlikely. We move past within ten metres of them and scramble over the fence.

Kholo Road, when we hit it, is kerbed. So often it's the small things that shock. The city is still ahead of us, but there's no denying we're in its penumbra now.

We are at the top of a ridge, and the river is now at the end of the road as it falls away to Kholo Crossing Bridge. But we've suddenly been thrust out into a world of written signs. At walking pace, each sign demands our attention, as if each, in that moment, is speaking to us alone. We are defenceless before them. I read them all as we pass, unable to tear my eyes away, signs on gates or hanging from fences, road signs, placards, advertising hoardings, posters.

Made in Australia. Taylex Tanks since 1969. Wastewater Treatment Systems & Rainwater Tank Specialists, with a platypus swimming around the words.

A plaque on a sandstone block: *Kholo Water Supply Pump Station. An Initiative of Ipswich City Council and Ipswich Water. Officially opened 5 October 2004.*

Someways further, a handwritten political slogan on corflute, capitalised in red pen. Ah yes, the federal election is today; democracy is scheduled to do its work. While John has already voted, I set out on this walk before pre-polling opened and so won't, for the first time in my life, be voting.

The last of the signs before we get to the bridge – a portable traffic sign standing confidently in the middle of the roadway – warns us it's closed. The first bridge crossing here was built in 1876. How many floods, I think, how many closures have there been, how many times has the bridge here been swept away?

But we don't need to cross the river, not just yet. That's tomorrow. Today our route keeps us this side of the bridge, and has us entering a vast riverside reserve controlled by the local water authority.

The bridge, as we near it, stinks. There are corrugations of slimy mud on the bitumen as the road takes its final curve down to the river. Then, fish. Hundreds of them, swept up and onto the roadway days ago, stranded at the high-water mark where the flood left them to writhe and drown in air, some already picked clean by birds. Heads and mud-filled gills, and gaping mouths, sharp combs of teeth and fin. As the rain continues to fall, the ribbings of mud begin to dissolve and a hundred silvered bodies start sliding slowly back down the bitumen towards the river. It's as if evolution itself is in retreat. As if the first prehistoric creatures to crawl out of the water to make their life on land are being dragged back to the watery depths, some colossal failure of creation.

We step over the fish, leaving our boot prints in the slime.

With the drop in water level over the last couple of days, the river is no longer overtopping the bridge. But the debris that's been left! There's a pedestrian walkway on the downriver side of the bridge. The railing has become a debris trap, with piles of upriver objects jammed against the rail, backed up, clumps of grass and leaf and stick and branch, suspended forever now, mid-eddy. Clear plastic bottles, chunks of styrofoam and black plastic buckets. Decorative baubles. Green triangular sapling barriers have been swept away along with their uprooted saplings, planted on the banks only recently to prevent erosion in the next flood. More fish caught in the debris traps or crowning mounds of grass and petro-chemicals, all washed downriver together, this their shared fate. Organic or plastic. Plant, animal or mineral. In the face of a river roused to anger, this is our shared fate.

~

We leave the swirl of debris and enter the reserve. The water management authority owns the land either side of the river, from the crossing here at Kholo all the way down to the water treatment plant at Mount Crosby. It's a bushland reserve, but not publicly accessible. It's bush as rugged as anywhere and yet so close to the city's edge, protected for water security reasons. Thick scrub, high basalt cliffs, creeks running off ridges and spilling into the river, into marshland. A number of times over the last year or so someone has mentioned the existence of a famed view from the top of the escarpment overlooking the river deep in the bush here, hushed longings, a fabled panorama. Few get to walk through here. We've been given permission to walk and to camp the night in here, and if we can find that escarpment …

We hug the river close until what we thought was a track begins to peter out. We don't want to stay at water-level and risk being caught between the water and the foot of the escarpment, with no way forward and no way to scale the cliff. We backtrack a little and cut inland. There is no route. But there *must* be a route – some four-wheel drive access track that follows the earth's bones as the land thrusts higher. We need to get higher. We crash through lantana, backtrack again, find clearings that close on us. The ground is marsh, and we sink. Welts on our cheeks become cuts. On a tree, we see a pink plastic ribbon – evidence that orienteers or surveyors have passed this way.

We cut across the earth. John's body, broken just two days ago, is powered by a purposeful energy. I hesitate at a gully, a patch of thick lantana on the opposite bank, its extent impossible to judge. Perhaps we should retreat and find an easier crossing? But John is learning to judge the possible from the impossible. He realises now there's actually nothing to fear in throwing your body at a wall of lantana, that after a struggle it usually gives. It's not just that there's nothing to fear – John's discovered a secret: that it's exhilarating. The give. The sweet sting of skin-scratch. The scent of lantana leaf, released by your own contact. And the wet – the steady falling rain, the chaos of droplets dislodged from bush and branch as we plunge forward, the pouring sweat. Our bodies are immersed in lantana. It's like swimming.

~

Again, the reality of what 'reserving' land means. Reserved from subdivision, from development, yes. From concrete and commercialisation. It's no small thing for a government to imagine a world with responsibilities more profound than revenue-raising. But reserved for what and for whom? Here it's water security, for all of us. Not unimportant. But has reserving this land caused it to grow wild? Become not only unpeopled, but unruly? Is that what happens when land is not carefully tended by human hands? That weeds have their way? Is this the primeval contest between the forces of order and those of chaos?

Are we stewards of the river, and all the lands through which it flows, and if so, what cosmological view guides us?

~

Eventually we find a four-wheel drive track. It has grown high with grass and is waterlogged, but it offers a way. Once upon a time someone perused this land – a logger or quarrier or grazier – laid out this route on a map, and then cut it out of the forest. Whatever its original purpose, it leads *somewhere*. It's the top of the escarpment we seek, and we will use this route for as long as it serves us.

As the track rises with the land it becomes a runnel. We lean into it, splashing our way forward up the slope. There is a fork, and another. Each time we stay left, taking the river-most way. The track we're on intersects a different type of pathway – a highway of powerlines overhead. We stop and look up and count the lines – fourteen – draped in perfect symmetry from the arms of a distant transmission tower standing silently in the mist like a many-armed Hindu deity.

In time the land plateaus and grows hard beneath our feet. Rocky, eroded, thinly soiled. Glints of quartz in the bone of the land. The forest is thinner too, sclerophyll, eucalypt. A tough treescape is softened by mist. But there is something else too, at the edge of cognition. The slightly unusual timbre of a sulphur-crested cockatoo's screech, a barely perceptible thinning in the canopy to the northwest, a faint shift in the bulk of the earth. Then we find what might be a footpath off to the left.

'Let's just—'

'Yes.'

The path rises gradually. Trees fall away, one by one, until ahead of us there is sky and beneath our feet rock, and the rock and sky draw us forward, impossible to resist, till the rock is no more, the earth is no more, and we are standing on a precipice, and if it is the edge of eternity, eternity is a river.

This river.

I look out. I look up. I look down. The river is wide, brown, swollen and flowing fast below us, flowing from the very horizon itself, straight towards us, before it turns away at the foot of the cliff far below and veers to its left. Watch it come. Watch it go. Feel its spirit. The river's presence is so great that all else disappears, including John, including me.

~

The dark pulls us away.

John sets the transistor radio centre stage on the floor of his tent. Both of us, at different times, have carried it here, with its black casing, grey tuning dial and retractable antenna. John pulls the antenna out as far as it can go. Hello little radio, I think. Bring us news of the world outside. John pours two fingers of Canadian Club whiskey into our mugs, mine plain silver aluminium, his chipped enamel with an ABC radio icon. The whiskey helps bridge the distance between a tent in the bush by a river and the three-yearly rite of democracy. Election booths have closed across the country, except in the west. Votes are already being counted. The national broadcaster's commentators are commentating. Soon enough the results will be known, and either a new government will be formed or an old one re-endorsed. Policy will change or stay the same. If a river could vote, I wonder, what would it say?

~

Some rivers do speak. Some have a voice.

Beyond the song of the Mississippi Langston Hughes heard in 'The Negro Speaks of Rivers', and beyond the whispering we all hear when we make love on the banks of a river. Beyond the ancestral spirits that speak to Indigenous peoples the world over. Beyond trade routes. Beyond fishing and swimming adventures. Beyond the roar of angry floodwaters. Beyond all that, some courts and some governments have said to some rivers: you have a right to speak and we must listen.

In 1972, in *Sierra Club v Morton* 405 U.S. 727 (1972), Justice Douglas was on the wrong side of the decision but the right side of history. If human societies could create something as artificial as a 'company' and give it rights, Douglas argued, then 'so it should be as respects valleys, alpine meadows, rivers, lakes, estuaries, beaches, ridges, groves of trees, swampland, or even air that feels the destructive pressures of modern technology and modern life'. Douglas was ahead of his time, but his vision is coming to pass.

In 2011, a provincial court in Ecuador recognised the Vilcabamba River – river of the 'sacred plain', altered by road widening, polluted by construction debris, narrowed, flooding – as having constitutional rights. Nature could be, and was, the plaintiff. A river could stand up for itself, through human agency.

Five years later, in Colombia, the catchment of the Atrato River – rising in the Andes, falling and flowing nearly 670 kilometres before emptying into the Caribbean Sea – had been gutted by mining and illegal logging, altering the very course of the river. Its Indigenous peoples suffered, and grew angry. The Constitutional Court ruled that the Atrato possesses rights to 'protection, conservation, maintenance and restoration' and that the river's rights are intertwined with those of its local communities, who see the river as a place of sustenance, a 'space to reproduce life and recreate culture'.

The Whanganui River – known as Te Awa Tupua by the Māori people connected to it – is another river recognised as having its own legal personhood. In 2017, New Zealand Parliament recognised the river and its tributaries as 'an indivisible and living whole, comprising the Whanganui River from the mountains to the sea, incorporating all its physical and metaphysical elements'.

In that same year, the high court in the Indian state of Uttarakhand granted legal personhood to the sacred Ganges and its tributary the Yamuna, which flow from the Himalayas – both great, life-giving, culture-nurturing rivers, but sick, choked by pollution, and in need of protection. Then, in 2019, in Bangladesh – a delta country, both flood-prone and flood-dependent, through which the Ganges passes and meets the ocean – the Supreme Court ruled that all the country's rivers had the status of 'living entities' and the rights of 'legal persons'.

This movement has reached Australia too. The Yarra, while not given legal personhood by the Victorian government, is still described in the *Yarra River Protection (Wilip-gin Birrarung murron) Act 2017* (Vic) as an indivisible living entity. The law recognises the 'cultural, social, environmental and amenity values of the Yarra River and the landscape in which the Yarra River is situated' and established an independent body to advocate on its behalf.

A decade of recognitions, springing from different cultural and historical circumstances. A decade of courthouses and houses of parliament – earthly determiners of rights – doing their rough work, attempting to give a voice to something numinous. It's a different form of corralling, this pressing rivers into legal frameworks. But the law can do no more than the law can do. And it is usually better for a legislator to speak than to remain silent, and for a judge to exercise sage courage rather than unimaginative caution.

But can parliaments and courts offer the recognition and protection that shared faiths and sacred myths once provided?

Not so very long ago, all human societies worshipped the deities of our rivers or lived with them in our dreamings or feared their divine personalities. Back when defiling a river was either inconceivable or such a breach of lore that the consequence was exile for transgressor and existential danger for its people.

~

John and I hear the voices on the radio. But listening to the election coverage is anti-climactic. Out here, beside the river, democracy has grown small. The democratic process is at work, a change of government is in the air, new policies, new faces on the national political stage. But none of the election commentators is resting midway through a river pilgrimage. It's a radio drama we're listening to, and it sounds tinny, confected. Despite the planet being in the midst of a sixth mass extinction and on the cusp of irreversible climate change, the urgency of the political moment feels, tonight at least, overblown. I'd thought the experience of walking the river, and all that the river had revealed about itself and our place in the world, might have prepared me to see the role of elections more clearly, might have sharpened their purpose. Instead, the election coverage is deflating, as if democracy is inadequate to meet the river's needs.

I ask John about covering the Indian Ocean Boxing Day tsunami of 2004, which he reported from Aceh, and what he's learnt about water's destructiveness. It is an ignorant question, simplistic, possibly even cruel. But John wants to answer. He's filed hundreds of thousands of words about this subject, spoken as many on air, had countless conversations about the flooding and its toll with colleagues. He could, here, now, reach for the words he's used before. He could return to the phrases he adopted during previous reflections. But he doesn't. He takes his time. The radio commentary continues in the background, but it is chatter. The river we have followed together for three exhausting days sweeps around us. The night too.

John looks at me. He wants, I think now, to do justice. Not to my question, but to another, the one art and suffering and rivers and night skies pose: who are we?

He shrugs. I shrug. We raise our mugs to our lips and drain the last of our whiskey, content, in this moment, for suffering and the poems we each carry with us and the stars behind the night clouds and the river we love, to do the talking.

Day 22

Mt Crosby Weir Nature Refuge
to Karana Downs

Distance: 15.72 km

Evening Camp: Place where the city begins

I return to the escarpment twice more: in the first drizzled light of dawn and then again as we're leaving – after I've stuffed my wet tent into my pack and my pack is heavy on my shoulders and the day's miles are yet ahead of us. I am careful in my looking, or try to be. I don't ever want to forget this place. I want to be able to describe this to my wife and sons, to my friends, to my sleepless self in my bed at home. So I set myself on a rock – the same rock as yesterday – and breathe in the contours of the place. The lips and ridges and horizons of the country. The way the river curves into sight upstream and then spears straight towards me, slicing through the thick forest as it approaches, aiming for my chest, the precision of its cut. And then the way it disappears again as it bends away downriver to the right. *Imprint that if you can.* I feel the vertical cliff-fall from where I sit. The river may be invisible immediately below me – so high is the cliff – but it consumes all imagination. *Close your eyes.* I do. *Now sketch what you see.* My hand moves through space. Sweeping brushstrokes, simple, like a child's. *Listen.* I turn my head. I hear the river. I hear the waters of a thousand creeks and gullies and tributaries. An entire catchment tips itself towards me, rush and rumble, welp and whisper. *Inhale.* The river smells of leaf litter and lichen and fish scales and soil and my sweat and that of all my walking companions. *What do you taste?* Change: someone's, somewhere, once upon a time, now.

~

We walk out of the forest. We reckon which track is ours, and reckon again when another crosses our path or where ours divides. We leap or wade across creeks, pass beneath a great Moreton Bay fig, hustle through long grass. We pass piles of gravel on the sides of the path, abandoned or waiting. The forest begins to thin. An electrical substation appears in a clearing off the ridge line to our right, so much muscular equipment performing so much electricity-splitting wizardry behind a chain-mail fence. We steer away, remaining within the forest for a little longer. The track descends, hardening, widening. Heavy machinery has come this way not so long ago. We pass beneath another highway of transmission lines.

Ahead a cattle dog. Further ahead its owner, running after it. John calls the escaped dog, and it comes to him. He holds it. Its owner reaches us and is grateful.

'Where have you come from?' she asks.

We explain how we'd entered the reserve yesterday, camped the night, and are only now walking out. She nods. She comes into the forest often to walk her dog, she says, but has never pressed too deep.

'Being a woman,' she says, 'even with a dog ...'

I don't know what to say. I stand before her, a man who has walked over 200 kilometres through the bush, without worrying about his safety, not really.

So I ask, 'Where are we?' She describes what is ahead of us. The road. The water-treatment plant. The river.

'Is the bridge open?' I enquire, meaning Mount Crosby Weir.

She pauses, then grins conspiratorially. 'It is for locals.'

The track meets the bitumen road, and there is no going back now. We turn left down the road and head towards the river.

The Mount Crosby water treatment plant is on the left behind a high fence, an impressive collection of warning signs hanging from the heavy retractable security gate: *STOP* and *Surveillance Cameras in Use* and *Authorised Access Only* and *HAZCHEM: In Emergency Call 000 Police or Fire Brigade* and *Caution: Keep Clear! Fire Ants.* Behind the fence is a complex of long rectangular pools and buildings and sheds and storeys-high water tanks. Behind the fence part of the city's water supply – including volumes recently arrived after a flood-propelled journey downriver from Lake Wivenhoe – is being treated for drinking. Organic matter, sediment, manganese, iron, other chemicals are filtered out. Chlorine and fluoride are added. Onwards the water will get pumped and piped. Opposite the treatment plant, on the other side of the road to our right, is a set of spill ponds, all brimming, a pair of black swans moving slowly towards us.

The river sounds ahead. It draws us on. A car passes us as we stride along the verge of the road, slows as it nears the river and the bridge. Stops. The driver remains in the car. After a few minutes, it does a U-turn and comes back towards us. John and I exchange nods with the driver – convention, a fellowship of sorts – before the car disappears behind us. Whether the driver has been thwarted by an impassable bridge or sated by the spectacle of a flooded river, it's impossible to tell.

The original purpose of the bridge was to transport coal from the deposits we walked past yesterday, across the river to fire the boilers to create the steam for the engines that pump water from here to the city. As the crow flies, the General Post Office is twenty-four kilometres away; as the river flows, it's closer to sixty-nine kilometres. What about the combined length of water-piping infrastructure supplied by this pumping station? But maybe that's not the question. The metropolitan part of what's called South East Queensland is serviced by a network of dams, reservoirs, treatment plants and twenty-two pumping stations, of which this is just one. Over six hundred kilometres of bulk water supply pipelines, tens of thousands of kilometres, perhaps more, of smaller pipage, leading from streets to buildings, then hanging suspended beneath bathrooms and kitchens.

The road curves and descends towards the river. A set of red-and-white portable water-filled barriers has been set up across the road to prevent vehicles progressing further. When the dog walker had said the bridge was open for locals, she'd meant, we realise now, local pedestrians. We slide between the barriers. Another barricade has been swung down and bolted into place immediately before the road gives way to the deck of the bridge. It might be adorned with an octagonal red stop sign, but if so, the river has made it invisible.

The bridge isn't merely a river crossing. Because it's not merely a bridge. It's also a weir, built in 1926. It holds the river back, agitates it, transforms it. In February the river overtopped the bridge. Today, while not flowing over it, the flooded river rams against the weir. The water is brown and flowing hard upriver of the weir. It comes out the other side – beneath the deck of the bridge – as white water, turbulent, indignant.

We climb over the barricade and onto the deck and immediately I feel the surging water vibrating through my body. I stand on the bridge and am suddenly vulnerable. I look ahead down the length of the deck to the other side. The crossing is narrow: a strip of single-lane bitumen laid down over the concrete weir, with two concrete verges and a pair of steel railings bolted to the concrete on either side of the ribbon of bitumen – but the downriver railing has gone, having been ripped off by February's flood. When the river tore the railing from the bridge, it also took chunks of concrete where the railing had been bolted down, leaving great bite marks the length of the deck, making the narrow crossing even narrower. What remains of the upriver railing is covered in debris – grass and reed and leaf and stick and root and branch. An entire tree trunk has been wedged against the railing and now leans, sleek and bare as bone, across the deck, a shuddering funerary arch.

I walk out onto the exposed deck. It's raining lightly. I don't want to slip, but still I walk quickly. Thousands of sharp sticks protrude from the wall of debris on our left, prodding us towards the unprotected side of the bridge. The weir and bridge rumble beneath our feet. The closer we get to the other side, the greater the damage to the deck. Even the upstream railing gives out and collapses inwards onto the bitumen, and the bitumen itself begins to break up, cracks becoming folds. The rumbling river roars ever louder. My blood pumps harder. I choose my steps carefully, will my feet into place, remind myself that locals and their dogs cross here. As we step off the deck and onto the road, the bitumen becomes so twisted and buckled that great heavy slabs of it lie loose, tossed one upon another. We step around the piled bitumen and over a collapsed chain-mail fence and are on the other side of the river.

~

It's my first time back on the left bank since Ian and I crossed to this side briefly at O'Shea's Crossing eleven days and 170 kilometres of walking ago.

Originally the plan from here had been to follow the left bank for twenty-odd kilometres down to Moggill and then to take the old cable car ferry back over to the right bank late this afternoon. But the advice from the ferry service is clear: *The river still remains too high to recommence safe operation of the ferry. We'll provide another update on Wednesday 25 May 2022.* That's three days away. Just for an update. With floodwaters still being released by the dam operators, the ferry won't be resuming any day soon.

I've resisted thinking too far ahead, have committed myself to responding to what's in front of me. To following the river and adjusting my route and plans depending on what we face, living in a rough and amorphous present, one that shifts its shape according to conditions.

But now that we've crossed over to Mount Crosby in Brisbane's outer suburbs, the organisational logic of the city forces itself upon us. Navigating a city demands different uses of time and space. The riverwalk's 'Where to now?' question becomes suddenly more immediate. If the ferry is down for three days minimum, what the hell am I going to do now?

John is leaving this afternoon. His wife is going to pick him up. But from where exactly? And now that we're in the suburbs, where will I sleep the night? And what of the planned seven-kilometre canoeing leg up ahead – the only way around the collection of prisons colonising the right bank of the river downstream from the Goodna boat ramp? Twice already I've updated my posse of fellow canoeists about the shifting itinerary, and the further the date for that paddling leg drifts out, from a weekend to a weekday, the more of them – including my son Liam, who's just started at a new school – can't make it. What will become of those seven kilometres? How? When? Who? It's almost too hard. Why can't I just keep walking?

If we keep walking, the next crossing down is Colleges Crossing. Perhaps it's open and offers a route back to the right bank – earlier than planned, sure, but that's okay, and there's a decent campsite over there too.

So yeah. Let's keep walking.

~

We lean into the hill, the river behind us, the red-bricked Mount Crosby pumping station to our right and the heritage-listed workers' cottages built here in 1892 a little further up the slope. At the top of the hill – the high bank – we look around and there can be no doubt. We are now entering the metropolis that is Brisbane City.

A metropolis that shares the name of its river is uncommon among river cities. There's San Antonio in the United States. Plymouth on the River Plym and Taunton on the Tone in England. Amsterdam on the Amstel, I guess – though in truth it was named after the dam. The Moskva flows through Moscow, and in Colombia, the capital city, Bogotá, is on its eponymous river. Given how dependent river cities are on their waterways, it's surprising that more don't recognise that debt in their names.

But maybe that's as it should be. Because a river's identity is separate to its city's. A river predates a city. Existed long before humans gathered on its banks, before language, before river sand was ever mixed into concrete. Rivers are shaped by a wholly different sense of time.

This river – the river I love, Maiwar – was renamed in the early 1820s for the New South Wales governor, Thomas Brisbane, a Scottish military officer with an interest in astronomy. It was a political gesture, entirely human, completely removed from any relationship the river had with the world around it until that point in time – its geography, ecology, the cosmology within which it was located. Many rivers in this country shared the same fate. The Karrawirra Parri (of the Kaurna people) became the Torrens in Adelaide, named for a British soldier and colonialist. The Dungala (Yorta Yorta) was turned into the Murray, in patronage of the British Secretary of State for War and the Colonies. The Barka (Barkindji) became the Darling after another governor of New South Wales. The Condamine in the Balonne catchment (possibly Mandandanji for water or running stream) was named for a mere aide-de-camp to Governor Darling.

Don't think naming – or renaming – is not radical. Don't think it is not disorientating, painful. Names matter. Might a different name generate greater respect, awe, new qualities of thoughtfulness? Love even? Might it cause us to pause? Might it help protect?

With each passing year, we drift further from Sir Thomas Brisbane.

~

After hugging the river so close for so long, this is a new world. Yet it is one I know well. We've entered the suburbs. If not exactly coexistent with city, suburbia cannot exist without city, is a manifestation of it.

At first it's a shock, and the shock is how orderly everything is, the result of overwhelmingly detailed planning. Planning which has become a template for how wealthy cities of the developed world are organised. Now, walking headlong into the cityscape after weeks in the countryside, its organising principles seem stark.

The streets are lined with poles strung with powerlines, which loop flowing electricity down to each house. There are cables for phone lines. Pipes for gas. Letterboxes are set into fences and on boundary posts.

We walk along a concrete footpath, comprised of segments with horizontal cuts to manage cracking. Either side of the footpath are grass verges, mown by someone whose responsibility it is to mow them. The footpaths run, in the main, perfectly parallel to the streets. The streets are laid with bitumen and are cut into the ground approximately twenty centimetres lower than the footpaths. There are concrete gutters either side of the streets. As a consequence, the streets become shallow canals when it rains. Some of the streets have rectangular slits cut into the gutters. These openings are too small for a person to crawl into, but big enough for water and rats and tennis balls. The openings are drains for the stormwater that runs down the street-canals when it rains. These ankle-height slits are all that is visible of a vast subterranean network of pipes that channels rainwater into the river.

Running perpendicular to the streets and crossing over the footpaths are driveways. Most are concrete. They lead from the street to garages and carports that are companion buildings to the houses that line the streets. The houses and carports are roofed with corrugated iron or tiles. Fringing every roof is a gutter, and each gutter is made of machine-cut metal of neatly uniform dimensions. The gutters gather rainwater and channel it down pipes bracketed to the sides of the houses. The downpipes themselves lead, eventually, to that invisible stormwater catchment system beneath the roads and footpaths. Some houses have water tanks, usually at the rear of the building, half-hidden. Those homes channel the rainwater from their roofs into the tanks first, and then, when full, channel the overflow onwards into the subterranean stormwater system.

There's a second network of pipes, separate from the stormwater system, also invisible from the street, that disposes of our water once we've used it – wastewater we call it – and which removes our own bodily waste at the same time. *Sewer*, from the Anglo-Norman *sewer(e)*, a channel for carrying off overflow from a fishpond.

A great deal of the organisational logic of cities is spent on water: how to get it to our houses, and how our houses themselves get rid of it. Strange that we don't keep all the water that falls on our own roofs and driveways, but instead pump different water in from dams and reservoirs tens of kilometres away.

~

As we approach Colleges Crossing, we are given fair warning: *Road Closed Ahead Due To Flooding. Reduce Speed.* Despite my sinking heart, we don't reduce speed but trudge steadily on to the crossing. The road bending down to the river is covered by a thick layer of muddy sediment. The water level has dropped, but still the water pours across the causeway, which is indeed closed due to flooding. A barefoot teenager with an umbrella follows us down to the river. He steps out into the mud, laying down a track, moving this way and that, a strange patterning of footprints, as if the mud is a canvas he needs to cover, and this is his signature, unique in all the world.

So, there really is no way across the river today. Possibly for days.

We retreat to a corner store back along the road to collect our thoughts and wait for inspiration. I order a serve of hot chips, and for John a pie.

'How's business?' I ask the shopkeeper.

'Slow. With the bridge closed, people aren't coming over.'

We collapse into aluminium seats beneath an awning outside the shop and slug cans of soft drink. Still the rain falls. We need to decide what to do. Or rather, *I* do. My feet begin to ache. I loosen my laces and then, after a minute or two, take my boots off altogether. John's waiting. The day has long ago tipped into afternoon. John doesn't want to leave me, but it's been four days, a day longer than he'd originally planned, and he has work tomorrow and he must go. He doesn't want to pressure me, but he should. I start running through the options aloud. There's a park ahead, Kookaburra Park. We should continue on that far at least. I can probably pitch a tent there for the night. It's only forty-five minutes from here, and John's wife can pick him up from there. But what about tomorrow? Well, I could continue on myself to the ferry the next day, a full day's walk along the left bank. 'Perhaps it will be operating again by then,' I say optimistically. Though we both know it won't. And if it's not, I guess I could keep walking along the left bank to the next bridge, the Centenary Bridge at Jindalee. But that is a barren thirty- or forty-kilometre route through suburban streets, away from the river.

'Why not paddle from Kookaburra Park?' John asks.

I think immediately of Steve's text from a few days back. He'd seen my update to the canoeing crew and watched as the lengthening walk upended the original plan.

In case it helps (no pressure), I'd be up for a long canoe day on the river ... starting finishing anywhere ... As the river is moving a bit, I think we could cover some distance. Bit of a wildcard option. Cheers, Steve

So, sitting barefoot at a rickety aluminium table outside a corner store, options closing in, I call Steve.

~

Yep, Steve's up for a day's paddling tomorrow but sensibly says he'll need to assess the condition of the river first. Though I've canoed before – a week on the Murray, different reaches of the Brisbane over the years, a stretch of the Balonne, a reach of the Condamine in a bathtub, up the tannin-leached Noosa – I'm an amateur. Once, as a kid, I needed rescuing when the outgoing tide near the mouth of the Maroochy grew faster than I could handle. I might be more proficient now, but I'm no expert. Steve, on the other hand, is a seriously experienced canoeist, adventurous but not stupid.

He meets us at Kookaburra Park in the late afternoon, while there's still light. We embrace. It's been seventeen days since we parted on the verandah of the Linville pub, and it's bloody good to see him again.

We all walk down to the bank together. Steve stops talking and starts inspecting the river intently, looking upstream then down, eyeing it off. I look too, but whether we go ahead or not depends on what Steve sees, not me. I glance at his face, once, twice, a third time, but it gives nothing away.

Does he see what I see? I see a fast-flowing, sediment-heavy river. I see eddies everywhere out there, small and large, close to the bank and midstream. There's not just one current in the river, but many, rivals braided together, competing, one swifter, another wider, one corrugated, another smooth. This is like no river I've canoed before. And the memory of the water's force this morning as it muscled its way over the weir at Mount Crosby comes surging back. But ... But there are very few sizeable pieces of debris out there – the river has largely flushed itself out. And the water level is dropping. And though it's still raining now – which means local run-off continues to flow into the river – the forecast for tomorrow is good. All of which I say to Steve, because whatever his calculus may be, I want to make sure all the positives are factored in too. And because, as I stand here, if the decision was mine to make, I'd risk it.

'What do you think?' Steve asks me.

'It's your call, Steve. Completely.'

John's loving this – both the gravity of the decision, and how it's being made.

'But what *do* you think?' Steve asks again.

I don't want him to think I'm cavalier. So I lay out all the negatives I can think of, including that the State Emergency Service has been warning people to stay off the river. There will be undercurrents, I say, and we don't know what conditions will be like further downriver. But I finish by returning to the positives, want them to linger.

Steve knows exactly what I'm doing.

'I agree,' he says. 'Let's go for it. See you at seven-thirty in the morning.'

~

John and I sit under a picnic shelter. It's still raining. He's rung his wife, who is on her way. Am I really going to put up the tent here? It's an open park, with nowhere secluded enough to be confident of an uninterrupted night. And the tent and my sleeping bag and my camping mat are all still wet.

'Donna could drop you off at a hotel for the night ...' John says.

But there are no hotels or motels anywhere nearby that are on the banks of the river. John feels guilty leaving me here by myself, but work and family call. So once again, there is the sadness of parting and nothing to be done, as John and his wife pull out of the carpark. Nothing to do but to feel its sharpness and to remember it is a blessing.

Dusk falls, but still people arrive at the park with or without dogs, to walk or just to gaze at the swollen, flowing river.

So I must wait. Until the gate is locked. Until it is completely dark. Until there is no-one left to come down to ask questions about a man setting up a tent in a public park. So I sit on the aluminium bench seat. The wind shifts direction and the gusting rain begins to spray over me. I change positions. The seat is wet and hard. My clothes are wet. Dark is falling and I begin to shiver. I am cold. I am cold and shivering, and before me is the prospect of crawling into a wet sleeping bag. I could do with a warm night tonight. I could do with the chance to dry out my gear, otherwise I'll be wet tonight and again tomorrow night. I pull my puffer jacket from my pack, put it on and zip it up. Still I shiver.

Though there are no hotels or motels out here, perhaps I can find a bed in someone's home? I log on to one of the room-sharing websites, Airbnb, and sure enough there are five hosts with rooms in riverside houses here at Karana Downs. But alas, all five of them are booked out tonight. Bugger it, I think, I'll send a message to them all anyway. Who knows?

Hi, I introduce myself. *I've been walking down the Brisbane River since 1/5 researching a book. I've reached Karana Downs and I'm looking for a place to stay, for one person, tonight 22/5, for 1 night. Might your home be available? Kind regards, Simon*

Four of the five hosts reply within minutes. Their messages are exceedingly polite, though their responses are expectedly, and understandably, the same: didn't you see that we're booked tonight? The night darkens. Vehicles intermittently nose down the road into the park, headlights catching me hunched in the shelter, a solitary figure seeking refuge from the drizzle with his backpack resting forlornly beside him. Couples with umbrellas climb out of their cars, those without stay within. I shiver. I'll have to wait until I am alone to put up my tent.

But the fifth host is different. Tanja is an artist about to set out on the Camino de Santiago. She cannot abandon one of her own. She tells me her partner, Joel, is a building-site manager who's on his way home from work and can pick me up in ten minutes.

~

Joel shows me where to leave my boots and gaiters up on his workbench and then sets an industrial fan in front of them for the night. Tanja points me in the direction of the washing machine and clothes dryer. I drape my tent and sleeping bag and air mattress over pieces of furniture, then lose myself in a warm shower.

A candle flame dances before us at the dinner table. The flame guides our talk, bending, flaring, guttering, rising again. We talk about men dying on building sites. About how many deaths a person might feel responsible for in a lifetime. We talk about gaps in the child protection system. We talk about spaghetti sauces and representations of death in David Bowie's lyrics. Because Bowie once bought one of Tanja's pieces. We talk about suffering and we talk about love – about whether we find it, or it us, and whether art might be a medium through which love passes. We talk about walking, and about coming and going. And we talk about the river, that too.

Tanja and Joel's place backs onto a creek that flows into the river less than a hundred metres away. Has the river ever been the subject of her work? Of course. But Tanja's not interested in flood and plenty, at least not right now. She's animated by absence, by fire, and by the conditions that make germination possible. I follow her downstairs to her studio.

And there the river is. Not in spate like it is outside tonight, but parched, gasping. Tanja's river is formed by two large sheets of paperbark, *Melaleuca quinquenervia*, placed beside each other, the left bank and the right bank, and the snaking gap between them is the riverbed. The edges of the paperbark banks have been burnt and become char-beaches. The bed itself is filled with text, a page from a medical encyclopedia upon which the sheets of bark have been laid. Words and half-words and phrases are visible between the river's scorched banks. My eye doesn't want to read; it wants to follow the curves of the river, to sweep across the page, reach by reach, to submit and be guided by the river's course, to follow, to flow. But this text does not flow. These words are static. I force myself to stop. To read. *ARTERIAL BLOOD ... may rupture, releasing ... into the abdominal ... causing PERITONITIS. ... Blood coming ... lungs ... the tissues ... oxygen ... relief of pain ... may be fatal ... dissolves easily ... swelling ... respiration ... collapse ...* On the right bank of the river, the bleached skull of a bird is fixed, its eye sockets over-large, as if it has been staring out at me too long.

~

In the midst of flood, it's hard to conceive of drought. Even though there was a bad one here just a geological heartbeat ago. Our memories are so attuned to what happened to us either yesterday or in our childhoods that we often struggle to remember what an approaching disaster feels like, even though we've lived through them before. Even attaching a grand name to the experience, like the 'Federation Drought' of 1895 to 1902 or the 'Millennium Drought' of 1997 to 2009, may not be enough. That's what stories are for, family legends, cultural myths, historical sagas. But while the catastrophe of flood is dramatic, the tragedy of drought creeps slowly and quietly.

Until it is here. Until, like in December 1895, the river runs dry – as it threatened to do again in 1902.

Last time around, from 1997 to 2009, the city grew increasingly concerned. Reservoirs dropped below twenty per cent capacity, water use was restricted, a desalination plant was constructed, the pipe network was expanded, prayers were uttered and eyes turned anxiously to the skies as Armageddon approached.

And yet even during the drought, twice as much rain fell on the city as it needed. During the drought, the city reused less than five per cent of the water it consumed. During that long decade, more wastewater and stormwater flowed out of the city than the volume of fresh water it used.

~

We are good at forgetting. We forget that we live, many of us, in a river. For that is what a flood plain is. We forget that from time to time the river will need to make itself known, and woe betide our forgetting. We forget also that from time to time the dry comes, dry upon dry, season upon season, accumulating like beads of bone on an ancient abacus. We cannot conceive that the water might slow. That a river might run dry. Until it does.

In my dream: the bleached skull of a bird with gaping eye sockets, staring at me until I wake.

Day 23

Karana Downs to Seventeen Mile Rocks

Distance: 36.64 km

Evening Camp: Place of the hidden rocks

Yesterday, today would have been unimaginable.

All that cloud, the rain, the mizzle – gone – and the sky is weeping-deep, relief-deep, wonder-deep blue. The flood has torn away half a park bench. I sit on what remains and look out. Universes hidden yesterday, fading memories even, are hinted at once more. A hawk fixes itself to the sky. The bough of a blue gum groans. The floodwater's retreat is writ clear on the sloping bank. Below the high-water mark, the grass is dead or dying. Wallows of mud. One by one the morning sun picks out drowned logs the size of men strewn along the floodmark. They would call out to each other if they could.

What is a rainbow but water and light?

~

Steve and his father-in-law arrive, a blue Wobbegong canoe strapped to the roof of their car. As he opens the door, Steve's eyes are already on the river. Is there an anxiety in him that hadn't been there yesterday? Perhaps his sleep has been filled with second thoughts.

Though maybe it's not anxiety at all, but merely intensity – because the long day's canoe journey we've planned is about to become real.

Steve doesn't take anything for granted. He treats me as if I've never paddled before, and I'm comfortable with that. We settle the logistics. We're aiming for Seventeen Mile Rocks, thirty-four kilometres downriver, and we'll call Steve's father-in-law to collect the canoe when we arrive. High tide near our destination, at Jindalee, was at 4.56 am, and low tide is just before midday. It's now a quarter to eight. In ordinary conditions, we'd be paddling with the current for four hours. But the floodwater throws out the usual reckoning of time and tide.

We wrap our essentials – food, drinking water, sunblock cream – in waterproof bags. I leave my pack and boots in the car to collect at the other end and slide on a spare pair of Steve's old sandshoes. We climb into our life vests. Steve hands me my paddle. Together we haul the canoe down the muddy boat ramp to the shore and load our drybags in the centre of the canoe's three cavities.

We put in on a quiet stretch of bank so Steve can assess my competency. I'm in the bow, he takes the stern. He demonstrates strokes: forward paddling, reverse, bow-rudder, J strokes, sweeps, draws. I mimic him. We talk about currents, and where the faster ones are usually found. He's concerned about eddies, and shows me how to navigate them. We do some drills. Once, twice, three times. Steve spies an eddy near the bank – shallow, harmless – and tests me. He'll do the hard work, make the judgements that matter. He doesn't say it, probably isn't even thinking it, but I hear it anyway: don't stuff up.

~

And then, after twenty-two days of walking, I am in a canoe on the river. Which means *in* the river, surrounded by it, carried by it. And bear us the river does.

255

The current takes us where it will. It feels almost predestined, this one-way downriver coursing. We paddle and pull along the river even faster, the thrilling, accelerating effect of our strokes.

There is a wholly different physicality to travelling the river like this. This is shoulder work, arm work, torso work. I am sitting on a moulded plastic bench seat, my legs extended in front of me. Bracing, leaning forward, digging my paddle into the river, pulling, rolling, leaning forward again, digging again. Half-a-dozen strokes on one side, swapping over, half-a-dozen on the other. Suck and rush, swirl and paddle-slap. Water running, dripping off blades. I feel my shoulders roll, the glory of it.

We are feeling each other out as well, the two of us working together, learning to synchronise our movements. Finding rhythm. Though it is Steve, sitting behind me, who is the principal guide, correcting the canoe's course with a deft stroke or by sinking his paddle into the water rudderwise. A flicked wrist. A shifting of body weight.

The river eddies violently, pitches us left, but soon we are through and running again, and there are new depths beneath us and the banks are racing past and perhaps we only imagined the river trying to throw us as it had.

~

But what is carrying us is not entirely natural. We are borne by flood release. This artificial flow. The dam operator's switch flicked in accordance with his water-release manual. We're riding that dam operator's tide.

Before he activates the spill, he issues warnings: to downstream property owners, to the councils, website updates. He flicks his switch and sits back in his chair, the weight of having to make a decision lifted. Or perhaps he leans forward over his monitor, anxious about what he has set in motion. The water comes faster out of the spillway than the steady build-up of a natural flood. Bridges may go under. Houses on the wrong side of the river may be isolated. Banks may erode.

The dam operator flicks his switch and the lake pours out of the dam and into the river, carrying a small canoe and its two small occupants.

We ride the flood release at more than six kilometres an hour.

We scan the river for debris. But most of the big stuff has already been washed out of the system. There is grass to contend with, sticks. We spot a log in an adjacent current, its path a little faster than ours. We are comrades, that log and the two of us in our canoe. Because what is a log but another river vessel?

~

Mangroves! Of course! I knew, but my thoughts have been elsewhere. We are in the tidal zone now, the mangrove zone, within reach of salt water, though still so far from the bay. The river is tidal to Colleges Crossing. It was not always like this. Once upon a time, the tidal reach was shorter and the river was fresher longer. The dredging changed everything. The dam too. After its construction in the 1980s, parts of the river that had been fresh water became marine. The regular high freshwater flows that used to flush salt water from the river dwindled. And after the dredging ceased, more silt and sand arrived on the banks, creeping upriver, mangroves following.

~

'When was the last time you canoed?'

Steve doesn't respond immediately, and then the silence becomes so long I think he mustn't have heard. I turn my head to throw my voice better.

'How long since you last paddled?'

'Ah,' Steve says, 'I wasn't going to tell you until we'd finished.'

'Nah,' I say, oblivious, 'tell me now.'

'Eighteen months.'

'That's a while,' I chatter on.

'Actually, the last time I was in a canoe, a friend drowned.'

I lift my paddle out of the water and turn around.

'My friend, J—.'

So Steve begins. It is a terrible story of rivers and their unexpected moods. Not this river, but one a few hours south, a river which is wide and slow and moves comfortably across a vast flood plain near its mouth. But inland, as it comes off the Great Dividing Range, it's wilder. Steve's lost count of how often he's paddled it, trips as short as an hour, as long as a week. It's a river he knows. Or thought he did.

He'd been paddling with one of his daughters in a group of four. He and his daughter were in a two-person Canadian. The other two were also extremely experienced, fit, water-hardened: J— in her seventies, and a man in his thirties, both in single canoes. The river was up and running hard, so they decided to paddle a calmer side-creek that day, and to wait twenty-four hours before risking the river. The creek was fine, better than fine, brilliant. They reached a long chute that ran straight for fifty metres before the creek bent sharply left and disappeared out of sight. They pulled in to discuss how they'd approach it. The man was a river-guide and the most accomplished of them. He'd go first and if there was danger ahead, he'd warn them. J— next, Steve and his daughter last. They confirmed the signals. A raised right hand at the bend before the river disappeared meant everything was fine. The river-guide pushed off the bank while the others watched. Before curving away, he raised his right hand. All was good. J— pushed off, canoed down the chute, disappeared left. Steve and his daughter followed.

The chute was swift, thrilling. Steve's daughter was in front, Steve behind. Father and daughter turned the bend and, Steve says, were confronted by roiling rapids. The water was thrashing wildly, and suddenly – desperately – they were battling to keep control, to stay upright. They swept past a perilous tangle of ti-trees on the right before they were thrown out of the canoe, and Steve yelled at his daughter to abandon it and swim clear, swim clear.

Steve shakes his head, even now, at how quickly things can change on the water.

He and his daughter dragged themselves to the bank, where the ashen river-guide waited.

'J—'s gone. She's drowned.'

They saw her then, trapped in the trees they'd skirted, partly submerged, her body caught in the churning water. They saw her while they tried in vain to paddle upstream to reach her. They saw her as they rang emergency services and waited for the helicopter to arrive. They saw her while emergency services advised them it was too dangerous to attempt to reach her. They saw J— all the sleepless night long, as the world turned and they waited for the water to recede. For the creek to relent and to allow them to recover her body.

'And so …'

But there is no need for Steve to bridge what happened eighteen months ago and what it's taken for him to get back on the water. And not just any water. A river in flood, another one.

We talk about J— as we paddle. I ask how often he'd canoed with her, what he knew of her family, what she was like. About her funeral. About what the last eighteen months have been like, for him, for his daughter, whether he knows how it's been for J—'s people. And we talk about how rivers just keep on bloody well flowing.

~

Steve has taken responsibility for me today, more than I – now knowing – would have wanted him to. But I'm glad he has. I hear the voice of Dostoevsky's Zosima from *The Brothers Karamazov*: each of us should accept responsibility, always, for everyone and everything. Family, friends, strangers, foes. The fate of a creek, a river, a lake, a planet.

~

That Steve's story should come now is almost too much, on the river, paddling our way towards Kholo Creek, another site of remembering – another woman, another death, also recent, but the circumstances so very, very different. Alisa insisted I go. It's the one thing she'd asked of me.

A woman who lived near here had been murdered ten years ago. A man killed her in the bedroom of their home. They had young children, girls. Then the man brought her body to this isolated creek in the middle of the night to dispose of it. He left her there and began to pretend. Pretended he knew nothing. That the woman had gone for an early-morning walk and not returned. The city searched, and still he pretended. People walked the streets, or kept watch, or scoured bushland. When she was found, it was by another canoeist, out on the river on an early-morning paddle, who just happened to turn down the creek to explore.

'You can't be so close,' Alisa said before I left, 'and not visit.'

We paddle slowly into the mouth of the creek. Two large concrete water pipes – a metre or more in diameter, backs stained dark with mould – run across it, one high, the other low, a pair of uneven parallel bars stretching from the water treatment plant to the city. On the other side of the pipes, looking down on them, is the road bridge.

There's no obvious site for us to pull in. The bank is a ribbon of mud three metres high. Above that is a dense patch of castor-oil plants, and beyond that is the road and the memorial. This won't be easy. It probably shouldn't be. We tie the canoe to the trunk of a submerged tree. We slip and fall and slide. We dig our fingers into the mud and pull our way up the bank. We grasp handfuls of mud and sapling and cut our palms on blades of grass. As the bank levels off, we wade through the castor-oil forest. Nearby, rising through the plants, is a tall metal pole, but it's only when we scramble up and onto the roadway that we're able to read the sign mounted at its top: *Please place flowers at memorial site ONLY.* We wait for a gap in the traffic and cross to a slab of pink sandstone, a photo of a young woman, Allison, in an oval frame, and a plaque that says: *After 11 days of searching, her body was discovered under this bridge.* There are flowers, yes, many, and all the rain these last weeks has dislodged leaves from an overhanging Chinese elm, a variegated carpet of foliage, too delicate for our muddy and tattered old sandshoes.

We climb back down to the water, where the killer disposed of Allison's body. He'd wanted her body to decompose and had called nature to his aid. The autopsy could not determine the cause of death, but ultimately his scheme had been thwarted. She'd been found sooner than he needed, and evidence of his crime remained with her body when it was discovered, leaves from her garden caught in her hair: seven crepe myrtle leaves, three cat's claw leaflets, fishbone fern, a lilly pilly leaf. Plants from her garden, rather than the banks of Kholo Creek. Crucial evidence to convict him.

We return to our canoe and push away from the bank. What will I tell Alisa about this place? I don't know yet. Perhaps it will be enough that I've come.

~

Another river joins us, the Bremer, a tributary merging from the right, from Ipswich, a city to the southwest. We aim for the point of land between the rivers, a she-oak at the very tip. The river is fast and seems to be speeding up. We nearly overshoot the landing spot and have to work hard to make it, then pull the canoe ashore, panting. There's a picnic table and benches trapped by mud. Back in 1864, one of the Petrie boys – among the earliest of the city's settler families – was awarded the contract to construct a training wall here, where the rivers meet. They were the days of river trade, a time before rail and road, when Ipswich's coal barges travelled downstream to Brisbane and beyond.

The she-oak leans downstream, pointing the way.

We put the canoe back in and follow the river down to the Moggill car ferry ramp on the left bank. We drag the canoe up the ramp's concrete ribbing. The twin barricades are down and chained together for good measure: *STOP, STOP*. Beyond the barricades, further up the bank, mud covers the approach road. The ferry is secured on the other side of the river. The words *Sea Link* – the name of the company operating the service – are visible from here, painted on its side in blue on gleaming white. The cables with which, in safer conditions, the ferry pulls itself along from one side of the river to the other disappear into the murk. Another set of cables crosses the river above us, powerlines.

Currents of air shape and reshape long strings of grey cloud over our heads.

We relaunch the canoe. There's a creek Steve knows. He wants to show it to me, or – perhaps – introduce the two of us. Six Mile Creek, at Redbank, narrow, dark, thick with mangroves. The water is still. We paddle slowly beneath arches of mangrove branches, the light dappling. We curve to the left, and the river, with all its noise, disappears behind us. We paddle through pools of golden light and dare not speak. A water dragon drops from a branch ahead of us, detonating against the surface of the creek, startling me. Steve, however, was expecting it, hoping for it, and he laughs. The first water dragon sets off a mad domino trail of falling, leaping lizards. Beyond the dragons are bridges, a pair of parallel highway and rail crossings spanning the creek, carrying commuters to and from the city, oblivious to this reptilian world below them.

~

We reach another watercourse emptying into the river, an enormous stormwater outlet with a wide, gaping maw. We paddle towards it, just as minutes earlier we'd paddled into the mouth of Six Mile Creek. But this tunnel is concrete, and straight, dark. I am closer, at the front, and peer deeper into it than Steve can. On the sloping inside wall of a section of the concrete pipe, stamped in black, is lettering never intended for public view – *Humes. Ipswich. Client/ Project* – and beside it handwritten notations in faded yellow and green. The bones of our urban water management infrastructure. I sense that if I suggested it, Steve would be keen to paddle inside the pipe itself, to explore it as far as light and boldness permit. I don't.

~

'A city is like a human body,' Steve says, 'the water pipes are your arteries, the wastewater-treatment plant your kidneys.'

'But—'

'Of course it's just a metaphor,' Steve says, knowing me well enough by now to wave away my instinct to test his analogy. 'But we're looking for ways to tell a story. Of cities where the use of resources is restorative, non-harmful and efficient, and allows humans and ecosystems to thrive. But it only works if you're capable of thinking in metaphor.'

He winks, I think.

'Our bodies consume food and water,' he says. 'We then use oxygen to convert that fuel into energy. We use that energy so our bodies can function, so our cells and organs can do their things. And we produce waste.'

'Our bodies are amazing,' he adds. 'They recycle water – reuse it – nineteen times, before discharging it as salty wastewater. While our cities discharge as waste ninety-five per cent of the water they suck in each day. If only ...'

Cities are metabolic. They may be heathy or sick. It matters what goes into them. They can be poisoned. They need to be looked after.

~

An enormous amount of work has been done to improve the health of the river. It's not true that politics doesn't matter, that elections are only performance. We can't allow ourselves to grow cynical. A lord mayor decides the river needs to be cleaned up. He's not the first. Others follow, at all levels of government. We turn towards the river. Sewage plants are cleaned up, rubbish traps installed, industry relocated away from the banks. Water quality is measured again and again.

Parts of the bank are revegetated. Only the hardest hearted of riverside property owners have not planted a tree on the banks of the river. Millions are planted to stabilise banks and maintain habitat. To trap sediments and stop nutrients leaching into the river. But which trees? Natives, yes, but which? Blue gums or she-oaks? Callistemons or lomandra? Is even that level of human intervention prone to unintended consequences? Might, for example, the leaf litter of some tree species have greater chemical concentrations of dissolved organic matter than others, so that overplanting of even a native species will detrimentally change the chemical composition of the soil? So that when the floods come and that soil is washed into the river, it might inadvertently poison the water? I think of John Muir's observation about trying to pick out anything in nature by itself and discovering that everything is hitched to everything else in the universe.

Who would presume to manage something as eternal as a river? Water alone might be manageable: its flow, and the volume of it, small or vast. It might be possible to direct water along pipes and into my kitchen, or the neighbourhood swimming pool, or a water reservoir. But a river is not a canal, let alone a pipe. A river is a universe.

~

'The river is really flying in that current a little to the right,' Steve says.

We veer, and run, and throw our heads back and laugh. What is that floating over there, half-submerged? We follow our curiosity that way.

A drone falls out of the sky and drops onto a mid-river island. We rescue it and return it to its pilot, Simon, a water-quality scientist friend waving from the bank. He's come to record our wayfaring, tracing us from above as we trace a river in spate. He changes batteries and follows our canoe downstream.

~

Entering the river from the right is another creek, Woogaroo, cutting through a high bank. The bank grows thick with vegetation, mangroves rising to eucalypts. Cat's claw vines drape and strangle and darken. Peering from a break in the foliage is a gathering of buildings, huddled close. This is a fabled institution, notorious, almost as old as the colony, of it and yet apart.

Wolston Park Hospital.

It has been known by other names since it opened in 1865 and its first cargo of sixty-nine people were shipped upriver from Brisbane Gaol on a steamer, the *Settler*. Woogaroo Lunatic Asylum, Goodna Asylum for the Insane, Goodna Hospital for the Insane, Brisbane Mental Hospital, Wolston Park Mental Hospital.

It's been known by different names, and it's held different people under vastly different circumstances. It was a place of treatment, yes, but sometimes also punishment. This was the final, lonely home of a great cricketer, Eddie Gilbert, born at the Durundur Aboriginal reserve on the banks of one of this river's tributary creeks, and the fastest bowler Bradman ever faced. It was also the home of children who were not psychiatrically unwell but were merely loved less than they should have been and labelled delinquents.

In a storage room of one of the art galleries on the south bank of the city, just downriver, is a painting I'd once stood before: *Passing the River at Woogaroo Reach* by Anne Wallace, from 2015. Nine women are adrift on a small boat on the river below the hospital. Seven of the women are standing; two are in wheelchairs. The hospital buildings are small, barely discernible, enveloped by lush forest. In the foreground, a snake steals an egg, a cormorant dries its wings, a torn mattress lies in a clearing and runes are scratched into the trunk of a eucalypt. These women, these 'delinquents', survived that place, Wolston Park.

I like the wordplay in the title – that what the artist and her subjects are trying to do is parse the river and that institution looking down at me now. As if understanding the wrongs of history might be within reach.

We lift our paddles from the river and drift, seeing those hospital buildings from the river for the first time.

~

Downstream of Woogaroo are more institutions on the right bank, these ones unambiguously prisons: two for men, one for women, one for children. The four jails within their secure perimeter fences are invisible from the river, set back from the water.

Children growing up in the suburbs on the bank opposite the prisons weave stories of escapees into their childhoods, and long, thin, uninhabited Cockatoo Island, hugging close to the right bank, is a character in these local escape legends. At night, they see muted campfires out there on the island. During the day the kids test themselves by canoeing across to the island, conquering their fear of escapees, eyes peeled for evidence of inmates' shelters – discarded prison clothes, cigarettes, caches of contraband, footprints in the mud. The kids smoke surreptitious cigarettes themselves and toss the butts, unthinkingly leaving evidence of prisoners for the imaginations of the next group of children who cross to the island.

Steve and I pull in and squelch through the mud and the island-long covering of castor-oil plants. We stand on this low mid-river island, water flowing either side of us. The only footprints are ours. The nearer right bank is thick with the trunks of clean white gum trees and cavorting undergrowth. In contrast, the left bank, on the other side of the main channel, is denuded, a serration of collapsing soil with piles of freshly eroded earth at the river's edge. Above us, clouds begin to gather once more.

~

Time and tide and flood release. The river propels us onwards but slows the further we travel as the influence of the incoming tide begins to counter the released floodwater.

The banks thicken with riverfront residences. Here there is wealth, beauty and occasionally the grace that money and taste can bring. The homes themselves are high and safe. Some of the banks are terraced by rockeries, others by sloping lawns. Most properties have paths or silvered walkways descending to private floating pontoons. Some of the pontoons survived February. Those that did not slid up their pylons on the rising floodwater – just as they were designed to – before popping off the top and floating away downstream, through the city, and out into the bay.

There is movement on the right bank, but what is it, and how much does it matter this time of the afternoon? But ahh! It's a glimpse of wonder, glimpsing us. A hind and her fawn watch us from the bank, where the creek at Mount Ommaney, deep in the city's suburbs, reaches the river. Over the years I've seen plenty of road signs in the suburbs hereabouts – *Beware Feral Deer* – but I've never spotted one this far into the city. Oh furtive little fallow, so you've made this home too. I am tired from all the paddling, from being exposed to the sun and the glare, and I hear the echo of William Blake's lines – *The wild deer wandring here & there / Keeps the Human Soul from Care* – and the doe and her calf looking at me from the bank are mythological creatures, and I have been touched.

But, little wild deer, you are a terrible pest, officially a 'restricted invasive' one. You are damaging the banks of that ancient creek with your hooves, ringbarking natives, spreading weeds as you feed. You cannot stay. You are marked, little deer.

Mother and child move behind the cover of bushes, and are gone.

I drift, setting Blake aside, to contemplate other creatures that make my soul leap. Echidnas, platypus, wallabies, racing emus, a mopoke on a midnight branch. But I cannot deny, schooled in Blake and Gerard Manley Hopkins and Dickinson, that deer do this too.

Onwards we paddle. Rivers are the major wildlife corridor through a city. A sea eagle rises from its nest atop a communications tower. We travel with birds and fish and bull sharks and on the bank all manner of creatures, wild and tamed.

We come upon a mannequin trapped in the branches of a large callistemon. She is pinned in such a way that she is upright, and knee deep in the water. Though her body faces us, her neck is turned slightly away and she gazes downstream. She is naked except for a red bra that has been pulled down around her waist by the force of floodwater. Her head is, in the way of mannequins, bald. Both her arms have been detached and are nowhere to be seen. Her smooth, pale surface is stained by mud and muck. She is only a mannequin, but somehow we can't simply paddle by.

We loose her from the branches that have held her captive and cradle her into the canoe. Perhaps she is sleeping. We take up our paddles once again and resume our journey.

The first of the city's traffic bridges is before us, the Centenary Bridge, crossing from Jindalee to Fig Tree Pocket. It is very high, with concrete under-ribbing. The afternoon is with us. Clouds begin manoeuvring into position for the evening. The bridge rumbles with peak hour afternoon traffic. The bridge's concrete footing is the largest human-built thing we have seen, up close, all day. We shepherd the mannequin through the widest channel beneath the bridge. Dark falls quickly. On the other side of the bridge, the right bank is a thin fringe of parkland for a kilometre to Seventeen Mile Rocks Riverside Park. We paddle close and look for a place to pull in. There is no avoiding a muddy disembarkation.

We shoulder our mannequin above the water, protecting her from the mud when we slip, and carry her to a fringe of blue gums above the mangroves. We set her down against a tree, seated upright. She is invisible from both the water and the walking path that leads to the park. But she watches us as we go back to the canoe – first for our gear, which we carry past her and lay out on the grass, and then when we return a second time to get the canoe itself, which we drag in laboured spurts up the bank. We have reached our destination for the day, Seventeen Mile Rocks Riverside Park.

~

We have passed innumerable jetties and pontoons and boat ramps today. We've passed the Mogill car ferry terminal and rowing clubs and hundreds of boats moored at private pontoons. On any ordinary day, we'd have encountered speedboats, canoes and kayaks. There would have been rowing boats – eights and fours and sculls – and jetskis and puttering tinnies. But today, nothing. We've paddled thirty-four kilometres across the course of an entire day and not seen a single other person on the water.

Tomorrow I resume walking. I cannot take the mannequin any further, and so we leave her on the riverbank, resting in the dusk against a blue gum's trunk.

~

The mystery of some rivers lies in their length. They come from distant lands, different from our own. Foreign tongues are spoken on their banks, they have different patrons and are protected by different spirits. Some rivers, like the Brahmaputra, carry multiple names along their lengths – reflecting different languages, cultures, mythologies.

We categorisers, we carvers of continents, we dividers of the earth into portions we can manage, have cut the river according to our view of the world. We legislate the management of land separately from water. The water authorities segment it differently again, oriented as they are to the management of water as a civic resource. This river is best conceived, they say, in three sections: the Upper River, from its source to the dam; the Mid Brisbane, from the dam to the water treatment plant at Mount Crosby; and lastly the Lower Brisbane, from the water treatment plant to the bay.

Settlers, too, travelling up and down the river by boat, divided it into 'reaches' as it flowed through the city; there are now thirty or more, the names usually coupled with the suburb through which the river passes. Steve and I have paddled past some of these: Daly's Reach, Moggill Reach, Redbank Reach, Goodna Reach, Cockatoo Reach, Gogg's Reach, Pope's Reach, Pullen Reach, Two Mile Reach, Mount Ommaney Reach, Mermaid Reach.

And if we were to continue together, there'd be more: Sherwood Reach, Chelmer Reach, Indooroopilly Reach, Canoe Reach, Long Pocket Reach, Cemetery Reach, St Lucia Reach, Toowong Reach, Milton Reach, South Brisbane Reach, Town Reach, Petrie Bight, Shafston Reach, Humbug Reach, Bulimba Reach, Hamilton Reach, Quarries Reach, Lytton Reach, Quarantine Flats Reach.

~

Steve's father-in-law meets us on the bank and we exchange canoe and sandshoes for backpack and boots. Now that we are looking to pitch our tents in a public park, it is *we* who don't want to be seen by others. We wait until dark, until after the picnickers and exercisers have left and the main gates to the Riverside Rocks Park are closed and the only way inside is by the narrow path snaking beside the river. We find a secluded site behind a stand of bamboo, on a bed of bark chips, directly beneath an elevated, rusting railway line.

After setting up camp, we wander back down to the river. Three beacons flash mid-river where a few shallow rocks remain, the last of the Seventeen Mile Rocks that once lay across the channel seventeen nautical miles from the city centre. The first of the rocks were detonated in 1862 to allow passage for boats further upriver – those coal barges from Ipswich and the ships filled with inmates for the Woogaroo Asylum. For a long time afterwards, a large cairn of terraced sandstone blocks warned boats to keep clear. But progress demanded further detonations. Intermittent removal operations occurred across the century. Despite it all, at low tide the currents over the remaining submerged rocks are a gnarled rush of torrent, and still the beacons flash: green, red and orange.

Beyond the beacons are the lights of the homes on the far side of the river. A chandelier hangs from a cavernous, high-ceilinged living room. We watch a couple moving around each other in that vast space, wondering how they will ever find one another. We follow the path downriver. A public place is transformed at night when it's emptied of people. It wants us to fill the space. Perhaps our spirits do expand. A low cloud settles over us. At our left shoulders as we walk is the river, less boisterous than during the day – it is now high tide, and the rush of flood release is muted by the incoming tidal flow. From ahead we hear another coursing sound, steady, undulating only with each subtle shift in the wind. I feel like I should recognise it, but after a long day paddling on the river – the sounds of the world being refracted off water or filtered through the space above it – it takes a while. Ah, yes, it's the noise of traffic on the arterial road in the east.

We wander down to the disused jetty of what had once been a cement works, and dangle our feet over the edge. The tide has turned and the dark river is quickening. Steve produces a small plastic bottle filled with port.

'J—,' he says lifting the little bottle to his lips, tipping back a nip or two.

He hands the bottle to me.

'J—,' I say after him.

Day 24

Seventeen Mile Rocks to St Lucia

Distance: 15.27 km

Evening Camp: Place of learning

Steve and I wake early and fold our tents in the dark, keen to avoid awkward conversations with park officers. We leave the park as the first early-morning walkers and their dogs begin to enter. Once again we farewell each other, the river journey having turned us both – at least for one night – into free-spirited guerrilla-campers.

I settle at a bench on the riverbank to wait for my sister, Kate. Ten thousand bluegum leaves are mirrored in the river's dawn, each serration. A half-submerged tree swims in pink, rocking gently in the current. Last night's beacons blink no longer, are now triangles of primary colour in a pastel river. A scrub turkey turns leaf litter over beside me. Does it conceive of river, does it perceive tides, is its life plotted by moons? It has risen, like us, to meet its day.

Yesterday's canoeing has propelled me deep into the city, thirty-four kilometres deeper. It's the furthest distance I've travelled in a day since the trip began. It did well, that little blue Wobbegong.

Canoes have, of course, been paddled along this river for millennia.

The next creek down from here, Benarrawa, now called Oxley Creek, had, for a short time, a third name: Canoe Creek. In those exchanges of names – Benarrawa to Canoe to Oxley – there are a thousand stories. Including one of the politics of colonial discovery, of reputations and claims upon history. Which means it's a story of power, influence and ignorance.

In late 1823, John Oxley, having been sent north by the governor to find a site for another penal colony, entered the river for the first time. On 5 December 1823, he wrote in his diary how 'amply gratified' he was 'in the discovery of this important River, as we sanguinely anticipated the most beneficial consequences as likely to result to the colony by the formation of a settlement on its banks'.

Yet while Oxley – Christian, naval midshipman, surveyor, ship commander, explorer, Sydney land-owner, colonialist, businessman, sheep breeder, company director, magistrate, diviner of penal colony sites – might have explored this river three times in total, and might have recommended the ultimate site of the city, he wasn't the first European to have located it when he arrived in Moreton Bay in November 1823. Captain James Cook, brilliant navigator though he might have been, hadn't found it either in May 1770 when he sailed the *Endeavour* past the island Mulgumpin, which he was responsible for renaming Moreton Island, after one of his patrons. Nor had Matthew Flinders found it in July 1799, when, accompanied by fellow explorer Bungaree, a Dharug man, he nosed the *Norfolk* into Moreton Bay, sniffed around and left again. Nor had John Bingle or William Edwardson found it on their inconclusive 1822 forays into the bay.

Sometimes the best time to find something is when you're not looking for it.

The first Europeans to find the river were lost. There were three of them. It was earlier in 1823. Sydney had been colonised thirty-five years before, and these three – Thomas Pamphlett, John Finnegan and Richard Parsons – were castaways and convicts, blown off course in a days-long storm, having set to sea south of Sydney on a timber-getting mission. So disorientated were they by the tempest that when, after twenty-two days adrift, they made landfall on Moreton Island, they thought they were still *south* of Sydney, and so set out northwards – the wrong direction entirely – on foot. The fourth of their group hadn't survived even that long, and was buried at sea.

That first whitemen's walk along the sands of Moreton Bay became a strange and pitiful odyssey around Kabi Kabi, Ningi Ningi, Ngugi and Quandamooka shorelines. For seven-and-a-half long months they island-hopped and riverbank-trekked. They were fed by the people of the bay, lived with them, travelled with them and received their counsel when they bickered and fell out with each other. They were given canoes, took others without permission, and fashioned yet others under the tutelage of First Nations guides. On Stradbroke/Minjerribah, the Noonuccal helped them cross to the mainland south of the river mouth and, ever fixed on heading north, the three began trekking northwards along the bayshore and reached the river. But they could not cross it. The mouth was too wide. They walked upriver for forty kilometres over the best part of a month, navigating tributaries much as I have these past weeks – walking up-creek until finding somewhere to cross, before coming back down the opposite bank when the creek joins the river again, to resume their river-trek.

By June 1823, they reached Benarrawa.

Benarrawa is wide enough to need a canoe to cross near its mouth and it happened that a canoe fashioned from stringybark lay on the far western bank of the creek. Pamphlett – the only swimmer among them – swam across to fetch it. And keep it. The three men then crossed the river itself in their stolen canoe. But the vegetation on the north bank was impenetrable. The men could find no northward path through the bush, so they retreated and returned back downriver, this time rapidly, in their pilfered canoe. Near what we call Kangaroo Point, opposite Meanjin in the heart of what is now the city, they found a second canoe, and took it too. As Pamphlett described it:

> … we fell in with a party of blacks, who were going to fish with their nets, and on our asking them, they gave us a good meal of fish; but the next day they seemed anxious that we should leave them; and upon our not doing so as readily as they wished, they made an attempt to seize our canoes.

'Our canoes': the three paddled away down to the river mouth and recommenced their futile northward trek. At Bribie Island/Yarun, Pamphlett and Finnegan grew tired. Or comfortable. They decided to settle with their Indigenous patrons and friends. Parsons – wild, mad, dangerous – continued north alone.

Then, on 29 November 1823, Oxley, in the *Mermaid*, spotted Pamphlett on the beach at Bribie, cooking the day's catch, and took him aboard. Finnegan was collected the following day.

'A river!' I imagine Oxley exclaiming as he interviewed Finnegan aboard the *Mermaid*.

Let me show you.

And so it was Finnegan who led Oxley to the mouth of the river, entered it again, this time in a whale boat, and led Oxley and his notebook up as far as the castaways had themselves already been, to Benarrawa, and a little further.

It was Oxley who named Benarrawa 'Canoe Creek'. A couple of years later, Edmund Lockyer, on his own upriver exploration, renamed the creek after his predecessor, Oxley. I wonder if, in doing so, Lockyer was hopeful that his name too might find its way onto the maps. As it ultimately did, Lockyer Creek. But there's an irony here. Perhaps, Lockyer mused aloud when he returned from his trip, the 'discovery' of the river shouldn't be attributed to Oxley – as the power-brokers of the colony had begun recording it – but more accurately to Pamphlett, Finnegan and Parsons. But the colony was not ready for truth-telling such as this. Oxley, it was determined, would be the river's discoverer. Truth-telling sometimes takes longer than it should. What stories might you have to share, Benarrawa?

~

'Coffee?'

I look up. My heart! I was waiting for Kate, but walking towards me, an almost too-beautiful surprise, is my mother too.

A map of a city holds pockets one knows better than others: homes and personal haunts, sites of joy and scar-tissue. Seventeen Mile Rocks is part of my personal riverside biogeography. This is where my mother lives, my three sisters. In this park, my nephews and my niece have had birthday parties. Here, this morning, Kate joins me to walk for the day and our mother brings us takeaway coffees. This place is threaded with family and relationship. This morning is infused with love. But Mum brings not just coffee: she wants to walk. She has two new hips, and one new knee, but she wants to join us, not far, not for long, but enough. Kate and I protest, but not too much – it would be futile. And so, for the first few hundred metres of the day, Mum accompanies us. Seeing us on our way, as she has done for a lifetime.

~

Kate and I must, this morning, walk across the high bridge I'd canoed beneath yesterday evening with Steve and the mannequin. Though there's a footpath for pedestrians on its downstream side, this is a car's bridge, part of the Centenary Highway, or, more brutally, the M5. There are four tight lanes, vehicles travelling eighty kilometres an hour and more. The early-morning traffic into the city is thick and desperate. Like the river, this route – the M5 – also leads to the centre of the city. But while the city was built where it was because of the river, this highway was built because of the city. The name of the bridge itself, opened in 1964, is marketing genius. Five years earlier, in 1959, on the centenary of the establishment of the colony of Queensland, real estate developers had the ear of government and council, and were given access to the western bank of the river for residential housing estates. The bridge was financed by the developers to funnel purchasers over the river.

Crossing the bridge on foot is an ordeal. Kate and I cease our talk – it's hopeless, the traffic noise is too loud. We bend our heads, wrap material over our mouths and noses to filter the fumes, and quicken our step. There's no thought of stopping midway to take in the view downriver. This is physical, urgent, a carbon monoxide and nitrogen oxide assault. Vehicles pour clouds of exhaust over us, mostly invisible now as the result of improvements in regulating emissions, but particles still spray from exhaust pipes, raining chemicals upon us. The bridge vibrates, not with the energy of the river passing through as the Mount Crosby Weir had, but with the weight of a thousand speeding vehicles. And it is hot: the burning fuel, the engines hot beneath their bonnets, the friction of tyres on bitumen and speeding car bodies through air. I'm enveloped by a cloud of rising greenhouse gas, a century-and-a-half in the making, and I'm sick and I'm responsible.

~

When we get off the bridge we fall into a riverside suburb where streets follow ridge lines. 'Leafy', these suburbs are called, houses set back from the road at the end of sloping driveways, hidden by trees. Eucalypts and jacarandas and poincianas and lilly pillies.

On the left bank of the river is a zoo, a city institution, fashioned as a koala sanctuary. It's a fine zoo, filled with koalas and eastern grey kangaroos and dingoes, a children's encyclopedia of Australian wildlife. The city's grandparents bring their grandchildren here. A city is known by the traditions it keeps, and those original grandchildren are now grandparents bringing their own grandkids. My earliest memory of a kangaroo is from here: clear, immediate, visceral. I remember reaching out and patting it – we're on a slope, the grass is cropped low, it's school holidays. It's the conquering of fear I remember, and the sensation of touching a kangaroo's stiff fur for the first time. But I took away another lesson too: that wild things can be tamed. An unintended lesson, no doubt – the sanctuary had opened in 1927 as a refuge for koalas being culled for fur during the Depression – but learnt subconsciously all the same. Some lessons take time to unlearn.

This is kangaroo and emu country. Koala country too.

~

Outside the zoo's perimeter fence, overlooking the river, is a meditation hut. Kate and I sit inside and unpeel muesli bars and rest awhile. Kate has rested in this hut a lot in recent years, meditating, practising gratefulness, looking out as the river goes by. It's a sacred place of sorts. A place where the distance between her cancer diagnosis and hearing a kookaburra laugh at a running river either disappears into nothing or expands into eternity.

~

At Indooroopilly, I wonder about the name. Everyone wonders about the name. *Yinduru-pilli* said Thomas Petrie. Probably a corruption of *yindurupilly*, meaning 'gully of running water'. Or perhaps it's a derivation of *nyinderu*, meaning leech, and *pilla* or *billa*, meaning creek or gully. 'Indooroopilly.' I say it out loud and wonder if there is a more beautiful, more perfect, word.

At Indooroopilly, flecks of gold were once found on the banks of the river. They came to nothing. The local cinema still carries the name of the dream, El Dorado.

At Indooroopilly, a new concrete footway curves out over the water. It passes, in turn, beneath a geometry of bridges. The river is gathering more of them the closer it gets to the city. We look up through the girders and the thick steel cabling. In the gap between twin railway bridges, the sky is purple and the slow clouds are a dense grey. The river reflects the bridge's piers. Kate and I pause to watch them change shape with the shifting sky, the moving river. Reflections are portals into different ways of seeing. And having entered that new reality, how can one ever be the same again?

Kate is strong. She is changed by cancer. Kate is amazing. She exudes gratitude. I am grateful beyond words to have this day with her. Kate and I continue on our way through the streets of Indooroopilly, this leech-creek-gully-running-water place.

~

We pass the tower bridge – give it its official name, the Walter Taylor Bridge. It's a suspension bridge, strung with forearm thick cables left over from the Sydney Harbour Bridge. The steel cables might connect two cities, Brisbane and Sydney, but fairytale towers like these connect us with our subconscious. There are two of them, twin towers, north and south. Each tower has rooms, beds, sinks, windows from which to look out at the lands below. People live up there, in the towers of the bridge, and have lived there ever since it was built shortly before the Second World War – bridge keepers, toll collectors, romantics, recluses. Kate and I walk beneath the northern tower and wonder about its inhabitants, as anyone who has ever passed this way has wondered. Are they like lighthouse keepers, trapped in their eyries by their singular duty? Do artists live there, and are those quarters their garrets? Is there a loom up there, a spindle and a sleeping woman waiting for a prince? Or has the tower been abandoned already by a cursed princess forgoing immortality so she can experience for herself the life of the city below, rather than merely gaze upon it through the reflection of a looking glass?

Kate tells me another story, which metamorphoses even as I hear it. A man once lived alone in the tower overlooking the river. He grew sad watching the world below, filled with its folly and ignorance and greed. As his sadness grew, so did he. The man and his sadness grew too large for the doors and the windows of the tower. When the man fell ill, the city came to tend him, but he was too large to be carried out the door. The city – itself always expanding – possessed miraculous machinery that it usually employed to help it grow. It used one of these machines, a great crane, to cut a hole in the tower wall and lower the man down to the street where an ambulance was waiting.

'What happened to the man?' I ask.

She doesn't know.

We walk quietly together for a while. I begin to imagine endings for the character I'm inventing. Liberated from his imprisoning tower, he found happiness among the people of the city. Or, the city tended his immediate wounds but could not see his sadness, and returned him to the tower lighter but no happier. Or, he found the city even sadder than he had perceived from his isolated tower and became overwhelmed by it.

But soon enough the river interrupts my fantasies, gently admonishes me: there are no endings, it whispers, only change.

~

Kate and I veer away from the river, where a fringe of waterfront houses forbids passage along its left bank. The sky shapes and reshapes itself, currents of air and masses of cloud. We cross a railway line and find a gathering of shops at an intersection. We doff our packs at an Indonesian restaurant and order gado gado. I ring Richard, ship's captain, about the final leg of the journey by his boat, the *Ballanda*, through the mouth of the river in a few days. The river has pulled the original date for the boat leg this way and that. Would he be ready to go in three days' time? He'll need to monitor the weather, he tells me. And the river level. And how much debris is coming down the river. But at this stage, he's right to go.

~

Kate leaves me at the entrance to the riverside college at St Lucia where I lived for my first two university years after moving to the city from Toowoomba. I'd contacted the head of the college, Greg, and asked if I could pitch my tent down by the river. Instead, he and his wife, Thérèse, offer me their guestroom. Is there no end to people's generosity?

I'm looking forward to seeing Greg and Thérèse but pause before going in. Carved into the sandstone block at the college's entrance are the words *Veritas vos liberabit*: the truth shall make you free. Behind the college is the river. I think of another Latin saying, Seneca's from the mouth of the king in his Oedipus, that truth hates delay, *Veritas odit moras* – if there is bad news to come, then spit it out.

The clouds have lifted, but the bad news is that this river will flood again. And the city will go under again, or parts of it. Forty-three of its 185 suburbs are riverside. To say nothing of all the creeks that wriggle through its suburbs into the river.

The city site-selectors knew, of course, about the dangers of flood, and looked for signs as they first gazed at this land with their European eyes.

Following his first expedition, in November and December 1823, John Oxley reported: 'There was no appearance of the River being even occasionally flooded, no mark being found more than 7 feet above the level which is little more than would be caused by the Flood Tide at High Water forcing back any unusual accumulation of waters in Rainy Seasons.'

But by his second journey, Oxley was growing suspicious. Kholo Creek had, he wrote, been 'at some period washed by an inundation: a flood would be too weak an expression to use for a collection of water rising to the height (full 50 feet), which the appearance of the shore here renders possible'. By 1841, we knew for certain: that was the year of the highest flood ever recorded. And all those that followed, 1893 and 1974 and 2011 and 2013 and 2022. We knew.

The First Nations people have always known.

We have built a city on a flood plain. And on the flood plain we continue to build.

~

Greg shows me around the college. There is a lightness in the halls I don't remember from the late 1980s. The mood of its corridors feels different from back then, and my first thought is that it's because the place is now co-residential, whereas when I was here it was a place for boys. And a place determined, it seemed to me then, to keep them boys. A place of strange initiation rituals on the other side of which too often lay not manhood but shame and resentment. Though that was a long time ago. See how capable we are of change?

Greg greets his students as we walk around the college grounds. A couple of girls tease him as they pass by, and we all laugh. Greg's easy generosity in the college is infectious. Though perhaps the late afternoon's buoyant mood is also explained by something much simpler and more universal: after weeks of rain, the sun is finally out.

Day 25

St Lucia to CBD

Distance: 13 km

Evening Camp: Place of the great steel bridge

Down at the college's pontoon, the river is full and almost still at the turn of tide. The day is brightening quickly and the images on the river's surface are sharp: the houses and jetties and trees on the opposite bank, the hem of upriver mangroves on this side, the tease of cloud, the deepening blue sky, my bearded face gazing back at me. The river is so sharp this morning you could shave in its reflection.

I look up from my image and see, across the water, the riverside houses of Yeronga. The name of the suburb is thought to be a Jagera word for sandy or gravelly place. When I'd lived here as a university student, I'd sat and gazed across the river at Yeronga as I read Jessica Anderson's *Tirra Lirra by the River* for an Australian literature course I was taking. It was one of the first novels I'd read about Brisbane. I began to see the river through it. In the novel, Anderson's protagonist returns to her childhood home in Yeronga after a lifetime away. She is old, the suburb has changed, and the river is now only accessible in a few places. She wants to walk by the river but can't find it. *Tirra lirra*, she hears the river saying, *tirra lirra*.

But the river is whispering to me now. *Tirra lirra, tirra lirra.* 'What about you?' it asks. 'You might have found me, but what do you see, what do you hear?'

Tirra lirra. They're not Turrbal or Jagera words, as one might think, but English – or not even English. An echo of a poem by Alfred, Lord Tennyson, 'The Lady of Shalott', inspired by Arthurian legend. '"Tirra lirra" by the river / Sang Sir Lancelot,' as the knight rode down to the river. It's the Lady of Shallot, not Lancelot, who is the subject of Tennyson's poem, a damsel who was cursed never to look directly upon reality, but to see it only in reflection. *Tirra lirra.* They're nonsense words, bird-song, brook-song. Meaningless, but impossible to forget. Words that whisper yet demand reckoning. *Tirra lirra.*

What do you see?

~

Today I walk alone. I am on my way to the heart of this city of a couple of million and more, but I walk in solitude.

An ant-bed track fringes the river behind some of the university's other residential colleges. I follow it, the river on my right. A collie with a white patch around its left eye falls in with me. A magpie carols. There is a fringe of mangroves on the shore line. Runners' footsteps approach from behind, crunching cadence of running shoe on sandy track. The dog's mistress calls and it obeys.

Past the colleges, up the hill to the left, is the sandstone university itself.

We call it sandstone, but we could just as easily call it waterstone. Stone shaped by the unhurried relentlessness of water's ways. Flooding rivers that deposited layer upon layer of sand – grain after grain of quartz – while time itself slowed. With each succeeding layer the weight bearing down on the earlier layers increased, the pressure of the newcomer quartz on the old-timer grains beneath.

Then, through the gaps between the grains, seeped calcite and silica and clay, cementing the quartzite granules together.

And *this* waterstone – this stone quarried for the grand facades of the buildings in this university's Great Court, the stone from which its grotesques and gargoyles are carved, and from which the zoologia fringing its buildings are cut – this stone comes from the river's watershed, up the catchment, where once, in the Mesozoic Era, there was a lake, 140 dinosaur-roaming kilometres from here. A lake turned into stone that now we quarry near the top of Lockyer Creek, at Helidon. Call it Helidon stone if we must.

The university's sandstone is alive with carvings. Friezes and figures leaning out to wink or grimace or smoke their pipes. Out of the stone leap famed academics or the coats of arms of fellow universities the world over. Above the geology and mineralogy building is a dinosaur frieze – creatures that lived and breathed while the first grains of sand in these stones were being laid down by the river. Also carved out of the stone are the plants and animals now living on the banks of the river: the leaves of the Moreton Bay chestnut trees Dominic and I camped beneath near Moore, the blue gum about to topple from the eroded banks of Lockyer Creek, bunya pine, small-leafed fig, eucalyptus galore, the grass tree from the forest at Andrew and Jill's, *tessellaris*, the ti-trees that have kept me company for weeks. All the river birds are there too: hawk and pelican and kookaburra and cormorant and duck and heron and eagle. And, of course, kangaroos.

~

My ankle begins to bite with pain. Ah, bugger, I'd forgotten that. And, now that I start thinking about it, I realise I am labouring.

The university's ferry terminal is closed for repair following February's flood. Should I have given myself another break to repair my ankle? There are lonely exercise stations at regular intervals between the path and the water. I stop at one to rest and swallow a couple of anti-inflammatories, and wait for them to kick in.

Across the river, at Hill End, long verandahs and decks reach into space, claiming river views. You could tell the history of a river city through its river views. It would be a history of beauty, of beholders and their eyes. It would be a history of private property, and land speculation, and reserving land for public parks. And, inevitably, it would be a history of dispossession.

I resume. I am in the mood for walking; the way is good, and I begin to forget my ankle. The sandy track gives way to concrete footpath. My legs take over. I think of the different paths I've followed these weeks: cattle pads, bandicoot and wallaby tracks through lantana, overgrown forest tracks, fire trails, roads, highways, verges of highways, grass footpaths, electricity wires, underground cabling, fence-lines, railway lines (functioning, abandoned, converted), concrete footpaths, satellites in their orbit, Alpha and Beta Centauri, the Milky Way. To say nothing of the river itself.

The set of four television broadcasting towers becomes visible on Mount Coot-tha to the west, one for each of the stations we've licensed to transmit news and entertainment to our televisions, reflecting back to us who we think we are or telling us who they want us to become. I had seen the towers yesterday from a different compass point while walking with my sister; then they were bunched like a stand of interlacing steel conifers. Now that they have fanned out along the ridge line of the mountain, they are loners.

There is no choice but to pull away from the river, where a fiat of houses commands the riverbank, and to veer up and onto a series of roads that leads to a shopping district called Toowong Village. There was once a creek here, and the area was known as Toowong, the prefix 'too' meaning 'sacred water'. But 'Village'? Sometimes developers name a project and the name sticks, no matter the poverty of its fit. This is a commercial and retail hub, not a village. There is a shock of cars and buses and shoppers getting in and out of their vehicles. A train line that disappears momentarily beneath buildings. The crisscrossing of local trade routes, an endless number of destinations for those of us with money in our pockets. I've shopped here often enough over the years, back when I was attuned to its rhythms. But now, everything about this tangle of shops is discordant, and I hurry to get through. Soon enough the riverside path reappears, and the strangling knot of commerce is behind me.

Still, there is debris. Is it fresh? Deposited by the floods of the last couple of weeks? Or is it rubbish from February, yet to be cleared? Though one piece of debris goes back even further. January 2011 destroyed a floating restaurant on this reach; it was called Drift. For over a decade, its skeleton remained moored to the bank, gathering graffiti and wasps' nests, no-one prepared to pay to break it up and remove it. It was an eyesore, yes, but it had become more important, in its death, than in its life. This is what flood can do. These reminders of powers greater than our own, these mementos of our mortality. But now it is gone. February lifted Drift higher, pushing it up and onto the pedestrian walkway, so that it was, finally, completely beached. Before I'd left the city, nearly a month ago, it rested askew across the path, impossible to navigate around, wrapped in warning tape. Now that it is gone, who will remember the damage a lazy river is capable of wreaking?

As I walk down Milton Reach, the city's skyscrapers appear in the distance, the corporate headquarters or branch offices of industry. The institutions of democracy lie up ahead too – Parliament and the courts and the government offices. The galleries and concert halls are all up ahead, across the river from the central business district, the CBD, on its south bank. The oldest of the city's buildings are also before me, the windmill on the hill and the old Commissariat Store on the river's north quay: memorials to where the city began.

The first of the bridges spanning the river in the CBD appears, a newer concrete toll bridge named by someone with an ear for a bad pun after a local popular music group, the Go Between Bridge. Other bridges appear. The Merivale railway bridge passing through a gentle archer's bow. The William Jolly vehicle bridge. The Kurilpa footbridge. Victoria Bridge between the CBD and South Brisbane. Just downriver from the Victoria Bridge, a new footbridge is under construction, long, white, tubular steel pieces, designed and located to lead pedestrians from the south bank into a new casino on the north bank opposite. Then another footbridge, the Goodwill Bridge, named for an athletics carnival held in the city in 2001 and sponsored by a media tycoon. The long, low span of the Captain Cook Bridge carrying the Southeast Freeway across the river is next.

Around the bend, not yet visible, is the city's great steel cantilever bridge, the postcard bridge, the Story Bridge.

What else is up ahead?

Echoes of the time before the city. Of Meanjin and its spirits.

Hear. Listen. They're not echoes.

~

The first time I heard the name Meanjin, I was a kid. It was the name of a literary journal, legendary even then. Current editions read: 'Founding editor Clem Christesen (1940–74). *Meanjin* was founded in 1940. The name, pronounced mee-an-jin, is derived from an Aboriginal word for the spike of land on which Brisbane sits.' A Turrbal word.

A few months after the colony of Queensland joined the Australian federation, in August 1901, a conversation took place in the pages of *The Queenslander* between two of its higher-profile white men, Archibald Meston, the 'Protector of Aboriginals', and an old Tom Petrie, famed for growing up with Indigenous kids and learning language. In 1901, Meston wrote:

In response to your inquiry concerning the aboriginal name of the Brisbane River. The Moreton Bay blacks had no generic name for river. They gave a name to every reach and bend, and every spot with which any remarkable incident was associated ... When on a visit to Brisbane, as a youth, in 1870, the old blacks gave two distinct pronunciations of the word. The mainland blacks called it 'Maginnchin', and the Stradbroke people 'Meeannjin'. Unless Mr. Thomas Petrie, now the oldest Queensland settler, and the best living authority for fifty years on the Brisbane dialect, can tell us the meaning of 'Maginnchin', then the origin is lost beyond recall.

The best living *white* authority, perhaps.

'Mr Meston,' Petrie replied a fortnight later, 'makes a mistake when he says "The Moreton Bay blacks had no generic name for river." The Brisbane tribe called river "Waar-rai" and creek "Yin-nell," but as Mr. Meston says, they had a name for every reach and bend, and therefore, as was suggested in another gentleman's letter, the natives never gave specific names to rivers.' As to the name for what had become Brisbane, Petrie writes, 'It was "Meeannjin" that the Brisbane blacks (not the Stradbroke people) called Brisbane, but at present I cannot remember the meaning; I will try, however, and think of it, and let you know.'

The absence of Jagera and Turrbal and Ugarapul and Quandamooka and Jinibara voices in that exchange is, now at least, startling. But the exchange reveals more than silence; it also lays bare the sudden need – the colony having dissolved, the history of a new nation being cast at lightning speed – for these state-makers to try to understand just what they were inventing. And it lays bare something else, another question: am I any less ignorant today than Meston a century ago?

~

There is an obelisk here, stunted, unprepossessing, wedged between North Quay and the Riverside Expressway, high on the north bank of the river. It is not beautiful. No flâneurs come here. No passers-by. It can barely even be glimpsed by car as drivers' eyes dart frantically left and right as they change lanes either to get onto the expressway or avoid it. The obelisk is as lonely a monument as has ever been erected. Which may be just as well. It's also a stark warning to anyone believing history might be pinned for eternity by inscribing words on stone: *Here John Oxley Landing to Look for Water Discovered the Site of this City. 28th September 1824.*

But Oxley neither discovered the site of the city – which has been here from time immemorial – nor did he even land here.

Oxley did land on the left bank on 28 September 1824, that much is true, on his second expedition up the river after Finnegan showed him the way the previous year. In the meantime he'd returned to Sydney to report his findings, that, yes, a new penal colony could be established at Moreton Bay. Go then, Sir Thomas Brisbane said to him, go and establish there a penal colony where the worst of the worst might be shipped, the recidivists and the unreformables and those among us who are more beast than human. So Oxley returned.

This second journey north was on a different ship, the *Amity*, and this time it was loaded with convicts – twenty-nine – and the inaugural commandant of the penal colony, an army man, Lieutenant Henry Miller, and the King's botanist, Allan Cunningham. On 15 September 1824, this motley community – convicts, soldiers, their wives, children – began to establish themselves on the Redcliffe peninsula out in the bay. But having reached Redcliffe, Oxley was keen to explore the river again. The very next day, 16 September 1824, Oxley and Cunningham left Commandant Miller and his fledgling colony in the bay and headed upstream, a second time, to explore.

Eleven days later, on Monday 27 September 1824, Oxley and Cunningham were on their way back downriver, having pushed on as far as the previous year, and further.

They reached a creek at sunset, near where Toowong Creek entered the river. The following morning Oxley's party decamped, and

> … proceeded down the river, landing about three-quarters of a mile from our sleeping place, to look for water, which we found in abundance, and of excellent quality, being at this season a chain of ponds watering a fine valley. The soil good, with timber and a few pines, by no means an ineligible station for a first settlement up the river.

The first *white* settlement on the river. That landing, that chain of ponds watering a valley, that good soil, became the 'X' on the city's first psychological map, worthy, in that way of things, of a foundation stone. But that wasn't here. Rather, those pools, that valley and the landing occurred further upriver than the spot where this lonely, traffic-strangled obelisk stands. There is nothing foundational about this spot, nothing to commemorate.

~

How does one love a city seeded, like this one, in violence?

The original settlement dream of Brisbane wasn't so much for a city as it was for a prison, with all its inherent cruelties. One overseen by war-hardened soldiers. Indeed, the city was doubly seeded by violence, because to build a settlement meant claiming land that was not theirs to claim.

When the convict settlement was relocated from the shores of the bay at Redcliffe to the banks of the river in 1825, better water was only part of the reason. The upriver spike of land chosen by

Commandant Miller for a fresh start was probably also selected because it could be defended more easily: the riverbanks were high, affording visibility and providing better defensive positions. Miller the soldier knew about besieging cities, and defending them. He'd fought with the 40th Foot Regiment in the Napoleonic Wars in Europe.

This is a city shaped by the experience of men who'd learnt about warfare in Napoleon's theatres. Not only were commandants Miller and Logan Peninsula veterans, but so was the governor, Sir Thomas Brisbane. All over the continent, soldiers like these were handed military or civilian duties. Thomas Mitchell and William Light and their theodolites. Charles Sturt with his river-tracing expeditions.

Napoleon's enduring, surprising, reach.

Downriver, out in the bay, there's an island, *Noogoon*, which became a prison for the new colony. But before it was institutionalised, in 1827 when it was an uninhabited island in the bay and the colony was itself still a jail, Commandant Logan exiled a Quandamooka headman named Eulope to the island for leading resistance activities against the colonisers. He was nicknamed 'Black Napoleon', and Logan called the island 'St Helena', after the island in the Atlantic where Bonaparte had been imprisoned, and where – as Logan was acutely aware – he had died just a few years earlier in 1821. Only Eulope's imprisonment lasted a mere three days. He 'simply stripped a sheet of bark, sewed both ends, and paddled back' to North Stradbroke Island, Minjerribah.

The genesis stories we choose, and those we choose not to choose. Stop, the river says. Listen. Think again.

~

It's a beautiful word, Meanjin, this land shaped like a spike. In planning the walk, a friend had suggested I talk with Uncle Joe Kirk, a First Nations Elder who'd written a book for children, *Duelgum*, about mother eel and her long Brisbane River journey. Where should we meet? I asked. At Meanjin, at the Botanic Gardens on the spike of land surrounded by river.

~

I turn my back on the obelisk and fall back down onto the riverside walkway and find myself in a Jeffrey Smart painting. A lone figure on a bitumen walkway beneath an elevated concrete expressway. Light slices through the geometries above my head, creates acute shadows on the surface of the water. Angles of light unknown on this river until these engineering feats began to be visited upon it ever so recently.

I pass beneath the Kurilpa footbridge. I walk through the shadow of Victoria Bridge, beside the new pedestrian bridge taking shape. Mountains of pale sand are being heaped on the reclaimed bank, tipper trucks releasing their great loads for bobcats to then nip away at the heap, bucket by bucket, builders taking what they need, spreading, mixing. This base construction material. How far has this sand travelled? I wonder. Perhaps, I think, it should have been dredged straight out of the riverbed immediately in front of this site. To sacralise the act of recycling.

~

Finally, finally, I reach Meanjin. The word has been acquiring new layers of meaning day by day, but feels lighter now, not heavier.

On Meanjin I sit with Uncle Joe Kirk at a table in the sun in the Botanic Gardens. Uncle Joe tells me about the sacred nature of water, of this river and its creeks, of names and history. The river is not snake, but freshwater eel, Uncle Joe says. Duelgum. Mother eel who travelled out of her waterhole to the river and then to lay her eggs in the sea. Uncle Joe tells me how the river was a boundary between the land of the Turrbal and that of the Jagera, and how they'd signal each other by fire to seek permission to cross. He points out where they'd cross at low tide, describes the shape of the river back then, less than two hundred years ago.

~

After Uncle Joe leaves, I return to the river and the path that follows its curve around the spike of land. There are moored yachts, and upriver ferries and downriver ones too. A sheer cliff on the other side of the water.

And then the unmissable bridge. The steel bridge of the golden age of bridge building: the Story Bridge.

This city is blessed with a bridge whose name transports us to a land of dream. Venice and its Bridge of Sighs. The Golden Gate Bridge of San Francisco. The Chapel Bridge of Lucerne. The Story Bridge over the Brisbane River. A story within a story: that a public servant's unremarkable surname might be bequeathed to a steel cantilever crossing, dressing it in mystery.

What story? Whatever story meets the need of the day. A story of linking a city's second commercial district – Fortitude Valley – with a southside shopping clientele. A tale of public spending to relieve unemployment following the Great Depression. One of parochial ingenuity, a bridge built 'by Queenslanders, for Queenslanders, with Queensland material'. Of quietly replicating the same design from another city, Montreal, half a world away, and forever linking those two cities. Ten thousand personal stories: of lovers rendezvousing beneath the bridge and gasping fireworks displays and accidents and suicides and notes to friends scrawled on the backs of postcards.

~

Alisa and I will be rendezvousing tonight, here, in the heart of the city. Inevitably, as I've been walking my way back to the city, I've been walking my way back to her. Across the river, on the right bank at Kangaroo Point, is a pub, the Story Bridge Hotel, an old three-storey place built in the 1880s that vies with another iconic riverside pub – the Breakfast Creek Hotel further downriver – as the city's most famous watering hole. It's a pub that's accommodated all types over the years. As a university student, I read poetry in the annexe behind the pub to an audience of fellow poets, curious bikers and random antagonists. Now, thinking about accommodation for the night, it's not a bad option. But I'm tired, and on the left bank, and in a couple of hours Alisa will be here. So I settle, instead, on a new place – also under the bridge – that opened in 2019 when the old wharf stretch known as 'Howard Smith's Wharves' was redeveloped.

~

Alisa and I share a drink in the hotel's rooftop bar. The deck of the bridge is above us still, its interlacing steel members sinking into the night sky. A wash of pink on the foundation arch. We look up, out. How the night glows, pulsing through bands of colour. The cantilevered bridge itself wears a necklace of light, pearls of glowing pink tonight, flares of gold. A low cloud absorbs it all, luminous with river and bridge and lovers. Through the arch, across the water, the ten thousand white lights of the city's office towers take their positions, layer after layer, too many to isolate. If they have secrets to tell, so do we all.

On the other side of the river, a giant poinciana has been illuminated by floodlight. A breeze moves through its branches; ripples of fireworks.

Day 26

CBD to New Farm

Distance: 12 km loop

Evening Camp: Place where crows land on high balconies

It is morning. The bridge is still above us. Outside the window the river has gone nowhere. Or rather, continues to go nowhere and everywhere. The river mouth and the bay are close now, a mere turn of the tide away. But the final leg of the journey can't begin until tomorrow when Richard and Desley and the *Ballanda* collect me from New Farm Park, just downriver from here.

Alisa has no time to linger – the demands of her day call her away – so I sit alone with a coffee and the hours ahead. I want to visit the art gallery at South Bank to see how the city's artists have depicted the river, but other than that the day is for wandering.

~

And so I begin making my way back towards the CBD, back upriver, where I plan to cross to South Bank by the Kurilpa footbridge. After following the river's flow for a month, it's strange to now be walking against the current.

I pass Customs House, this version of the customs office dating to 1886, though the river trade had been regulated from this site for decades before then. It is a grand building: two levels, a wide frontage, rendered brick on a stone foundation, double stone stairs rising from jetty-level, all crowned with a copper dome. There are also two entrances – one from the river, and the other on the opposite side of the building, from the street, Queen Street – projecting the authority the fledgling colony's leaders sought for the institution tasked with collecting its taxes.

I re-enter the Botanic Gardens and thread my way between the temporary public art installations that populate the park this time of year during the city's *Botanica* festival. The crowd is thick around the edge of the largest of the gardens' ponds, a hum of talk and gesturing and heads turning to companions. I wait my turn and shuffle to the front. In the middle of the pond, an artist has erected a huge-scale replica of the Story Bridge, the bridge half-submerged by its waters, drowning in the waters of a mega-flood. The aesthetics are stunning: water and steel and light and colour and reflection. There is an easy symmetry, but the tensions in the piece run deep. If this is a premonition of catastrophe, there is also a prayer here. For beauty. And for art.

~

I navigate the skyscraper-shadowed streets of the CBD. It feels like I'm in an eddy. For years I've walked these streets. I've walked them purposefully, performing errands or chasing trinkets. I've mooched down them, buoyant or blue, losing myself. I've walked alone and in company. Pushed prams and trolleys and pushbikes with punctured tyres on these footpaths. Broken into jogs. Shivered and sweated. I've lost and found much in these alleys. But never have I been here like this, entering them after twenty-six days on foot with the river as companion and guide.

All around me is concrete. I've learnt to see it on this walk and cannot unsee it. This environment we are building. All these materials that come from somewhere. That *must*, if we want to live as we do, come from somewhere. We extract sand and gravel and rock from riverside quarries and from hill sides. From the bed of the river itself, from its ancient alluvial plains.

All around is cement.

All around is gravel.

All around is sand.

Everything is water.

I pass between towers of river.

~

Eventually I stand on the Kurilpa footbridge, the sun at one o'clock, the masts and spars of the bridge set like a brig upon the river.

The river below me is brown and wide. I wait, watch, begin to see. There is always something to see. The surface of the river begins to move under a gentle breeze. Suddenly I see the rays of sunlight, reflecting off the corrugated currents, break into bursts of brilliant light. It comes as revelation. Stars on the water. I fix my gaze on one as it bursts into scintillant being, sparkles, then disappears, the duration of its glittering existence no longer than the time it takes for its ripple of wake to collapse and be replaced by another. A second or two of life, no more. But that new ripple contains its own star, and the next ripple another. Ten thousand stars, each so briefly lived, together a galaxy of fireworks, trembling with wonder. Stand on the bridge and take the time to watch. The gallery ahead will wait, wants you to wait. This moment. These stars, this river, each breath of life.

~

Outside the gallery a bronze water rat, a kuril, eyes off a bronze elephant, flips the poor giant onto its head, marooning it on the

299

bank of the river. The kuril turns away. I bend to read the plaque by the sculpture. *Traditional Elder Uncle Des Sandy tells how the kuril is intrinsically linked to the mangroves that weave around the Kurilpa Point shoreline, which feed it and provide it with shelter, and that these trees, with their strong tentacle-like roots, are the source of nourishment for a diverse ecology.*

In the art gallery hang half-a-dozen paintings of the river. There are grand, city-building paintings. *Panorama of Brisbane*, Joseph Augustine Clarke's four-metre long vista, had been commissioned by the colony for the International Exhibition in Melbourne in 1880–81 – the event that would inaugurate Melbourne's Exhibition Building. This was the era of extravagant international exhibitions, of cities prancing and promoting themselves, since London's Crystal Palace Exhibition of 1851 – of civilisation itself straining to prove its glorious worth.

After Clarke gave voice to the young settlement's claim for recognition, Isaac Walter Jenner painted Brisbane on the banks of the river in the late 1880s. In *Brisbane from Bowen Terrace, New Farm*, the river traffic is increasing, ships and rowing boats competing for water space, the river filling out.

By 1897 it was Brisbane's turn to host an International Exhibition, another opportunity for fresh representations of the river and the city. George Wishart's *A Busy Corner of the Brisbane River* is dominated by a multiple-masted cargo ship being unloaded at dock. It was important to show the world that the city was busy. And that it had corners worthy of examination – the time had passed when a single canvas could claim to represent the whole of the place. Wishart's sailing ship achieved something else too: its spars and rigging inspired the design of the Kurilpa footbridge I crossed an hour ago.

And there too, is Anne Wallace's *Passing the River at Woogaroo Reach*, no longer in storage, hung in all its strange magnificence. Telling its thousand stories of endurance and transformation. Of all the gallery's river canvases, this is the one I owe most to: I've experienced the river differently because of it.

~

The sun is beginning to fall. It is time to go. I seek out the boardwalk. It is good to be walking with the current once more, back downriver, towards New Farm, where friends, Peter and Myra, have offered a bed for the night at their riverside apartment.

Two seagulls perch on the railing between path and river. Their orange webbed feet are brilliant against the dark grey railing, their pure white breasts cut sharp against the river surface behind them. I had always thought the long, low concrete arches of the Captain Cook Bridge to be pure white, but I see now that – compared to the gulls – they are grey and stained by vertical rainwater runnels. The harder I look, the more the bridge becomes striped. A ferry passes beneath it, low and sleek. I look for the name and find it, the ferry I'd been hoping to see: the *Neville Bonner*, after the Indigenous senator. I'm glad. James – whose woven lomandra twine memento from the first day of this walk is in my pack – is one of Neville Bonner's great-nephews.

The walkway passes along the foot of the Kangaroo Point Cliffs, burnished and glowing in the late-afternoon light. This once-upon-a-time quarry, cut first on Commandant Logan's orders in the early days of the penal colony, had once been a rocky ridge, rather than the vertical cliff it now is. The rock made its way into some of the colony's oldest public buildings, into ship's ballast, into roadworks and, for decades, into the river's expanding system of training and retaining walls, until there was no more rock to quarry and the site was retired – having been, to use that uncomfortable phrase, 'worked out'.

Now rock climbers rigged in harnesses and ropes move up the cliff-face. The more experienced or cavalier move swiftly across the porphyry, while others lurch from hold to hold, pressing themselves against the face to catch breath. One of the climbers has been stationary for a long time, his foot wedged deeply into a crack. He turns to look across the river towards the Botanic Gardens in the crook of the river's bend, at Meanjin. What does one see when one is frightened? Exhausted? And how much distance from a thing does it take to know whether a change is evolution, revolution, or merely change?

~

I am joined by Peter, who has finished his work for the day. We stretch our legs past the remnants of the Naval Stores slipway and boatshed, outpacing a small log floating downriver on the outgoing tide. We come off the river, and when we reach the Story Bridge Hotel, climb the steps up to the bridge's deck and roadway.

The noise. The traffic. It is just after five. The day's light is dying fast now. The current of traffic is three lanes thick each way and every vehicle has its headlights on. The railing between the road deck and the pedestrian footpath seems slight. We take the footpath on the downriver side of the bridge. The traffic is coming towards us, headlit engine after headlit engine. If there are people in those cars, they're invisible. The engine noise threatens to overwhelm us. Or rather, me. The deck rises. We share the footpath with cyclists who coast downhill towards us. We walk up the rise, towards the steel superstructure. The deck rumbles. Joggers thud past. *Who cares about you?* a sign asks. *We do! Call Lifeline 13 11 15.*

~

We come down off the bridge.

The rising riverside towers cast shadows over the water. Rivers in modern cities routinely become canalised, but what prevents those rivers also becoming *canyonised*, with walls of high-rise buildings crowding out both banks? What, Peter wonders, stands in the way of a future where the river becomes a mere channel upon which the sun rarely shines directly? I think of the relationship between river and building in other cities – Paris, London, Vienna – whose waterside buildings are low-set, a handful of storeys high, no more. Brisbane erects imposing towers and allows them to fringe the river. People who reside or work in them look down on the river but are barely connected to it. I have worked in such towers. You can see the river – and it is beautiful – but you can't hear it, or smell it, or dip your toe into it.

Is this a complaint about where Western civilisation has led us? Or a more local one about the vision for this city? About how much shadow a city can cast upon its river before it is no longer a river?

~

Peter and I pull up stools on the balcony of his seventh floor apartment in New Farm. We look out across the river, southeast. A late crow lands on the railing and tilts its head at us. Yes, bird, thank you. Here we are on the balcony of a shadow-casting tower enjoying a lovely view. The crow leaves. We look out without speaking. In time flying foxes begin to emerge from their roosts in the mangroves and the eucalypts on the banks of the city's creeks, and follow the night where it leads.

I look out at the city we are building, a city on a flood plain. In this moment I can't help but think it's a miracle of sorts. A work of art as much as engineering, of imagination as much as planning. A city composed of river. A big brown snake-shaped city. A city groaning at times under the weight of our passion. For it, for ourselves, for each other. And straining too under the weight of history. Because it is in cities such as this that we seek ourselves, exhaust ourselves, make mistakes, atone, fail to atone, find love, give it. And where there is love, there is the chance to see anew. In cities like this we record ourselves and we make ourselves. We seek what is sacred, find it, are guided by it. Don't we? Haven't we always?

Day 27

New Farm to Moreton Bay

Distance: 25.63 km (24.63 km – 13.3 nautical miles – by boat)

Journey's end: Place of the saltwater bay

There is a breeze, cool, coming off the water, reaching even up here on Peter's balcony.

Out on the river is a single sailing boat. Its mast is undressed. It is at anchor, its anchor chain taut. I watch the lonely boat and feel a growing sadness. Today is my last day on the river. Today I will walk along its bank to the old power station, where I will join Richard and Desley and together we will take his boat down the river and out through the mouth to the shallow salty bay. And then we will moor, and then I will go home. I think again of Basho's observations about partings: that the one who goes is sad, and the one who remains grieves. But it is not always possible to know if one is leaving or being left behind.

I watch the little sailing boat in the river pulling against its anchor chain for what seems like a long time. But then something happens. The little boat begins to move upriver. It travels in small graceful sweeps, veering left, then dipping its bow ever so slightly before turning back and swinging right, lowering its bow again before turning back once more, jigging its way elegantly up the sparkling river. I can't make out the boat's name at this distance and lean forward, as if that will make a difference, some need to attach a unique name to this extraordinary feat. This pilotless, anchor-dragging boat, somehow liberated, and travelling free up the glittering river, transcending the bounds of time and place. Oh, see the river glitter! See it dance! Yes, I suddenly realise, see it dance. Because it is the river that is moving, not the boat. It is the river's surface that is alive – this play of tide and breeze and light. The river travels through time and space and form while the little sailing boat remains anchored. By the gauge of the apartment block on the bank behind it, the boat has not progressed a millimetre. But the glittering river allows it to fly.

~

I leave the apartment and make my way along the riverside path. I move slowly, a little stiffly. My pack is slung over my right shoulder. I still carry my walking sticks – retracted – in my right hand. I probably won't need them, but they've been loyal, and it's become a habit, and if people think they're an affectation I don't care. I pass other sailing boats moored off the bank. People will live in them for months on end, or otherwise simply moor them there. An inflatable police boat skips up the river, moving more quickly than any other vessel on the water.

Adjacent to the rose garden at New Farm Park is a ferry terminal. A city council bus with its distinctive yellow and blue colours is parked at the end of the street, its nose facing the water: *Not in service*. A crane with a long orange-and-black neck reaches out over the water as it swings its load from one side of a plot of land to the other. I hear a plane above my head, but it is faint. The crow in the poinciana tree at my right is louder than the engines of the 737.

The New Farm power station once generated electricity for the city's tram network, but the coal trains long ago stopped carting their black rock from the coalmines further up the catchment. The powerhouse has been reimagined into a great performing arts space. Though this morning it's not the transformed powerhouse I'm making for, but the newly inaugurated floating pontoon in front of it. A 'river hub' it's called, with four fingers for smaller vessels to berth, and an outer face dedicated to larger tour boats.

~

Beside me, as I wait at the café by the New Farm pontoon for Richard and Desley and the *Ballanda,* is a sculpture in huge red metal lettering: *f l o o d.* Only the top portion of the word is visible, most of it submerged by the concrete footpath, just enough to make out the word, but large enough for kids to clamber over. In my memory and in my imagination the table at which I am sitting was completely submerged in 2011, and near enough in February.

Two cyclists dismount from their racing bikes, hang their helmets off their handlebars, unpeel their fingerless gloves and tuck them inside their helmets before taking the table beside me.

'The most rainfall we've ever had in May,' one says.

'Just the weather system, innit?' the second replies.

The topic dries up. They move on to something safer.

~

The river flows quickly. Water from the deluge is still making its way downstream, still playing with the usual tides here. The clouds are high and in the east and distant. The cormorant perched high on a pylon on the opposite bank has no intention of leaving its post. A mullet jumps. I turn my head, but not quickly enough. I see only ripple. As I watch the rings expand on the surface of the brown river, the *Ballanda* appears around the bend, moving steadily upriver.

The *Ballanda* is a 33-foot timber-hulled, single-masted 'cruiser', launched in 1968, the year of my birth. Its hull is gleaming white, with neat yellow trim. There are three portholes on either side of the bow, worn like twin sets of onyx studs on a neat dress shirt. The foredeck is timber with long thin strips of weathered teak, and the mast rises from the deck just in front of the cabin. Aft of the cabin (there's a lexicon for boats that a stranger to them only ever uses clumsily) is a shaded, cushioned bench, and a set of free-standing chairs. Richard has fashioned his own flagstaff at its stern, and has hung a timber name-board to the boat's transom. The *Ballanda* is a thing of beauty, and it is loved. But it's also been designed for its environment. Its draft is just three feet, which makes it perfect for Moreton Bay's shallows, a 'Moreton Bay Cruiser', a boat whose builder is forever paired with it, part of its story, this one a Bert Ellis.

There was a time, not so long ago, when the river teemed with vessels. Now most of the boating activity in the river takes place near its mouth at the port. Once, vessels of trade and war, of pleasure and purpose, inhabited the city reaches and further upriver. Brisbane was obviously a port city, and the river was central to its maritime workings. Now there are few boats in the city centre: the regular passenger ferries, the odd barge or tug on some very particular errand, and the yachts and speedboats and dinghies moored at their private pontoons. Many are used so irregularly that they are less boat than adornment.

Richard and Desley don't moor the *Ballanda* immediately. They come around once to check conditions, Richard craning his neck from the cabin with one hand on the wheel, Desley leaning over the starboard side before moving across to port as the boat swings around.

'Ahoy there!' I call.

Desley rolls her eyes, then smiles.

'Simon,' Richard says sternly in response, before breaking into a smile as well.

The tide is no surprise. High tide here at the New Farm pontoon had been at 7.45 am. It's now midmorning, and we'll have the tide with us as we power downriver. But the river is full, no doubt about it, and the river is flowing fast, though not so fast as to abort the trip. What they're really looking for is debris, but there is nothing, or nothing large enough to worry Richard.

~

These last hours on the river. I want to greedily examine the banks for clues to the river's story. I want to ask Richard – heritage architect, bearded old salt, Brisbanite – questions as they come to me.

'Why does the river widen so quickly?' I ask as we pull away from the Powerhouse and make our way down Bulimba Reach.

'For shipping.'

Of course, the wharves.

'What type of dredge is that?'

'A suction dredge.'

'Who was it who built Newstead House, again?' It being the oldest surviving home in the city, at the mouth of Breakfast Creek.

'I think it was Patrick Leslie, the squatter, in 1846, but it was Captain John Wickham who you might be thinking about.'

Yes, Wickham, crewman on Charles Darwin's HMS *Beagle*.

'What's the maximum speed of the *Ballanda*?'

'Five-and-a-half, six knots, but with the tide we're doing eight-and-a-half, nine.'

When? Who? What? How? Why?

I want to learn all that he knows. Now! There's not a moment to waste. But I'll drink in too much too quickly if I'm not careful, and will see and hear and remember nothing. Desley smiles at me. She knows. I love Desley. I leave the cabin and step onto the foredeck and breathe deeply.

Hello, river.

I close my eyes and feel the movement of the boat on the river: engine, tide, current, flood release, the wake of other boats.

Hello, river.

~

We are heading for the bay. We are going slowly – not much faster than the speed Steve and I had paddled the river just four days ago – but we are travelling far too quickly. I want to savour this, but in no time we'll be at the mouth, will reach the bay. Time expands and contracts. Time, that current that bears us from birth to whatever horizon we are destined to reach. How quickly time runs today. There was another time, not so very long ago at all – eighteen thousand years ago – that the bay was not a bay, but was land, a plain. Back then the coast was twenty-five kilometres offshore. Back then – during the Ice Age – the river was narrower, steeper. And then the ice melted, and the sea rose, and the great plain of what is now Moreton Bay was flooded. Geologically, that was yesterday, just six thousand years ago. What was a plain is now a shallow bay, and a boat like the *Ballanda*, with its shallow draft, is made for travelling through the river's mouth and navigating this young bay.

A mouth that at the peak of its sea-rise flooding six thousand years ago was in a very different place, from Nudgee and Nundah in the north across to Bulimba on the south bank.

~

Stories begin to pass me by on the banks. So many 'once upon a times' and 'this was the place where' and 'since time immemorials'. Now is not, I realise, the time to learn new things, to acquire new facts. It is time to recognise, to remember, to retell.

On the left is Breakfast Creek. Can a single creek foretell so much about the history of a place? First, whose breakfast? Oxley's, in 1824 on his second trip up the river, this one with the botanist, Allan Cunningham.

So, Breakfast Creek. The creek of Dalinkua and Dalipie and Oxley and the legend of Harriet the Galapagos tortoise and of Darwin.

And of the *Ballanda* too – Richard has the boat repaired at a slip on Breakfast Creek. He tells me a story.

'See the flagstaff?'

Now that I look at it, it is striking: caramel-coloured with knots so dark they seem charred. 'The owner of the shipyard had to cut back the mangrove that was growing over the slip, so I asked whether I could have a piece.'

So this boat specifically designed for sailing in shallow Moreton Bay now has a flagstaff hewn from the mangrove growing on its river's shores.

~

After Breakfast Creek, the river bends right and widens even further. We pass the riverside suburbs. Hamilton on the left, Murarrie on the right. Mansions giving way to apartment complexes giving way to factories, industry.

Richard explains how the instruments on his boat work. The measurement of depth and speed and direction. He has different versions of charts in the cabin. Clipped to a board is a paper chart and beside it a screen showing our position on the river in real time. The river is blue in the charts, but Lockyer Valley dirt brown outside. Another instrument sounds the depth below us, deep in the river channel.

We pass beneath the twin Gateway Bridges, parallel high concrete. Between them, when we look up, the sun is behind a spoil of cloud. Pinkenba. Hemmant.

~

Hello, Coal Wharf, over there on the right bank. I have to say I'm surprised to see you here. But you're hard at work still, aren't you?

I can see your black coal shining from my midstream vantage point. Coal pours from your chute onto the ground, piling high. Your river barges ceased their passage long ago, but coal trains still find coalmines, and coalmines clients. Coal Wharf, your time has passed.

There is, finally, the Port of Brisbane. I've flown over you before, mighty port. Look at you now, from river level, all your towering cranes and gantries, all your vast yards of perfectly replicated shipping containers, so neatly stacked. I recognise the names of your clients too – Maersk, Evergreen, TS Lines, Cisco, Yang Ming, UASC. I've met them in other ports around the world over the years, on the backs of river barges and on trains speeding inland from coasts. They're not strangers to any of us. And look at you as you expand ever further out into the bay, as tip trucks loaded with earth and rock dump piles of land into the water. You call this 'reclamation', as if you are retrieving something that had once been yours. But that's not really what's happening, is it, hungry port? You're not re-taking are you – just taking?

Or as the poet Samuel Wagan Watson put it, 'artificial land with its artificial spirits, and the luck that floated here, with nothing to guide it'.

~

The river is now so wide, and growing wider. Its banks are receding. What is distinctive about the river's landscape withdraws from sight. Gums, mangroves, little rusting jetties, fishermen, lovers walking hand in hand. Over to the left, is that a tributary or a crane's shadow? Perhaps even a cross-current, answering some unfathomable law? I'm finding it harder to discern what I know is unique about my river.

This might once have been an estuary. Too much has been dredged now, too many banks artificially retained, walled.

Imagination and memory eddy. My river threatens to merge with other rivers.

A blue plastic barrel bobs its way downstream. Is it fanciful to imagine that one of the very cargo ships we passed a moment ago at the port had originally carried that plastic barrel to these shores from a factory in Vietnam or China or the Philippines? For a flood to then bear it back downriver to the port again, past the merchant vessel that had been part of its original journey here, past the port and out through the mouth towards the boundless Pacific.

A seagull follows us. We are a small boat, tiny beside those cargo ships, but big enough to leave a wake. The seagull finds nothing and flies away. Beyond our wake, back upriver in the distance, the towers of the city are still visible, rising above the dark fringe of mangrove.

'*Ballanda*?' I ask Richard.

'A derivation of "Hollander". So, "white man",' he answers.

Up ahead, through the mouth, is the bay. This is where these twenty-seven days were always headed. This is the end of the river. Out there is soil from the Lockyer Valley. Out there is river sand, deposited after millions of years of rock-grinding and flooding. All manner of things have been swept down here in flood: pontoons, logs, boats, cattle. The flotsam and jetsam of city life. If Steve and I had not picked her up, this is where the next flood might have carried our mannequin.

~

The river mouth is deceptive. Identifying precisely where the river ends and the bay begins is difficult, perhaps impossible. And ultimately a useless endeavour. Not everything can be measured. Some things resist exactitude. The glittering of the sun on the surface of a bay. The materialisation of clouds. The ways of rivers. The meaning of dreams. A soul's journey.

'The coffee pots,' Desley says, 'look.'

Desley points towards a three-legged beacon up ahead to the right. The legs rise out of the water to a platform and railings.

Erected vertically from the platform is a long rectangular cage, crisscrossed with diagonal strengthening members. Atop it is a small solar panel and a gauge of some sort. I could ask her what the gauge measures, but it doesn't matter. Not today. Desley waits for me to take it in, before pointing to another beacon, off to the left, another tripod and platform. This one is shorter, and the body of the beacon comprises four bright red-painted metal flanges, set at right angles to each other.

'When we pass through them, we'll know we're in the bay.'

They're not coffee pots, of course. They're channel markers through which Richard is steering the *Ballanda*. Markers, also, of the particular relationship those who know the river at its mouth have with it. A maritime language evolved to this river, this time, this place. I think of the farmers three hundred kilometres upstream, and the phrases they have adopted, its 'freshes' and its 'holes' and its 'runs'.

A snaking river sheds its skin.

~

Then we are through, and the coffee pots are behind us and we are in the bay.

This is salty water, shallow but wide and generous, and in the distance is the sweep of the long sand islands sheltering the bay from the east – Minjerribah and Mulgumpin, and further north, Yarun. Nearer are the smaller islands that dot the inside of the bay, including St Helena, or Noogoon, where Eulope, 'Black Napoleon', laughed and swam his way to some form of freedom. Passenger ferries make their way to the islands and the sails of the racing yachts are full. Here the dugongs feed in the seagrass beds and seagulls gather on sandbanks at low tide and the sun's arc is long and the horizon is always beyond reach.

The river – bounded, banked – has gone.

Epilogue

I lay out, on the kitchen table, the objects I gathered on the walk. James' woven lomandra leaf is there. The colour has gone from his twine, but it is tougher for the time that has passed. A small pair of deer's cast-offs. Half-a-dozen riverbed pebbles: green, blue, red. A longer, steely-grey stone that Steve and I had hoped might spark, but which never did. A mussel shell.

I'd originally planned to collect one object each day as a way of documenting the changing environment of the river across the length of its journey. But after a couple of weeks the exercise began to feel forced – a chore. I found myself arriving at camp without yet having found the day's object, the demands of navigating the terrain in the rain more immediate and more pressing. Exhausted, often bedraggled, I'd then have to explore the campsite's surrounds with the pressure of needing to locate an object that represented the day's walking. I was trying to squeeze the river and the walk into too rigid a frame. It was artificial.

So I abandoned the obligation of daily memento-finding, and waited instead until an object drew me to it; for the river to issue an invitation. Some of the objects I gathered were startling, others mundane. But each, in its own way, had been part of the river's call.

What do these objects mean now, months later? I pick them up, one by one, and examine them, feel their textures, smell them, remember, wonder.

There's a rusted steel railway sleeper pin I picked up with Dominic, heavy now in my palm. There is the bark of a bloodwood and of an ironbark, still damp. A sprig of dried callistemon leaves. My Noogoora burr is there, somehow smaller than when I'd fingered it for days in the pocket of my rain jacket. A fluff of cotton that Ian had teased in two, half for him, half for me. A small chunk of bitumen dislodged by flood from a submerged highway. The tooth of a cow that had come down to the river to drink or die. All that cattle country. A crow's ash seed pod, with its five spiked fingers, is part of the collection too. A segment of bark from the floor of the grass-tree stand I'd passed through with Grantley, and with it a black-bean pod, its dried seeds rattling around inside like a musical instrument. There's the snap of blue plastic prised from the weave of flood debris pressed hard against the railing of one of the flooded bridges John and I had ventured onto. Finally, a tiny white shell, which I lift to my ear. Is that the sound of the river I hear? Or is it human song?

I set the shell carefully back on the table.

So these are my river mementos. The river is in these objects, but they are not the river.

Where is the river then?

It is just over there, a short walk away, making its way through the city as it flows by.

~

Last night I dreamt of the river. Though in truth I couldn't tell if it was my river, some other river, or all rivers. As I woke, the dream-river became the gutter outside my bedroom window, and I could hear a sharp-clawed possum making its way along the gutter's tin edge. And then, as I lay and listened, it began to rain. The possum's clattering ceased as it sought shelter beneath the eave. The rain grew heavier. I could picture the possum sitting on its haunches, looking out, waiting for the drumming rain to pass. I tried to picture the drops as they fell, tried to isolate them, one from another, until the effort pulled me into an eddy of outlandish thought, wild, but oddly comforting. I thought that perhaps, just perhaps, among all the drops falling on my roof in that steady shower of rain was a drop that had fallen upon me once before. Or, if not an entire raindrop, then a single molecule of water. That from that shower was a molecule that had fallen upon one of my companions or me as we walked along the banks of the river one wet May.

Is that so strange a notion? That water flows through one's imagination as surely as it passes down a river or a gutter? A fanciful thought, yes, but one with the power of myth. Perhaps it's true.

Acknowledgements

This book – and my walk along the banks of the Brisbane River in May 2022 – was propelled by a great tide of generosity.

The river and its country have been here for a long, long time. I acknowledge with deepening appreciation and gratitude the Traditional Owners of the lands through which it flows, and who have cared for it for thousands of years: the Traditional Owners of the Jinnibara, Ugarapul, Jagera, Turrbal and Quandamooka lands. In particular I want to thank Jason Murphy and the Board of the Jinibara People Aboriginal Corporation, James Bonner, Maroochy Barambah, and Uncle Joe Kirk for their words of encouragement, advice or knowledge sharing.

To those who shared the miles: Steve Kenway, James Bonner, Dominic Cleary, Ian Cleary, Grantley Smith, John Taylor, Kate Cleary, Sue Cleary, Peter Jensen, Richard Allom and Desley Allom-Stewart – they were unforgettable miles, weren't they?

My great thanks to those who offered a riverbank tent site or shelter including: Marjorie Martin, the Lions Club at Camp Duckadang, Tracey Diver and Tanya Grimward at the Linville Hotel, Kelvin and Joan Allery, Catharina and Stephen Kusay, Sharelle and Paul Cooper, Alan Roughan, Michelle May and

Charles Brabazon, Jason and Laurisa Wendt, Andrew and Jill Stallman, Josh Olyslagers, Anne McDowell, SEQ Water, Tanja Stark and Joel McGuin, Brisbane City Council, Greg and Thérèse Eddy and Kings College, and Peter Jensen and Myra Poon.

Many people gave extensive guidance or shared their knowledge of the river and its ways. I am particularly grateful to Graham Brown and Andrew Stallman for their logistical support and encouragement – if I leant too heavily and too often on them at times, they didn't seem to mind. Enormous thanks to Jos Bailey, Donita Bundy, Simon Costanza, Cheryl Gaedtke, John and Bev Harris, Ray Kerkhove, Anne McDowell, Noel and Helen and Ken Schmidt, and Daniel Scott who were unstintingly generous with support and advice.

Others who have left an indelible mark on this book are Rob Adair, Chris Andreas, Damien Atkinson, Robyn Brown, Christy Clark, Nell Clarke, Margaret Cook, Tony Cox, Louise Cullinan, Jim Darcy, Patrick Dixon, David Fagan, Robert Ferris, Piet Filet, Terry Fitzpatrick, Caz Gardam, Renae Grace, Jennifer Harrison, Mason Hellyer, Shona Jackson, Mario Lattanzi, Bruce Lord, Nicholas Loos, Chrissy McCoombes, Cam McKenzie, Alida and Ian McKern, Samantha Mills, Peter Munro, Graham Orr, Seamus Parker, Mary Philip, Felicity Plunkett, Ben Ponte, Lyle Robson, Christine Sammut, Jim Soorley, Ian Townsend, Andrew Travis, Jeff Tullberg, Anne Wallace, Maya Ward and Darren Zanow. Thank you.

I thank all those people up and down the river who live with it and were immediately and so generously supportive of this project, including: Veronica Albury, Kelvin and Kylie Allery, John Annan, Peter Baker, Steve Barraclough, Jason Bell, Russell Bernitt, Nick Bischoff, Robert Bishop, Andrew Blinco, Graeme Brown, Mervyn Burow, Peter Carseldine, Brenden Christensen, James Christensen, Scott Cleary, Peter Cooper, Mark Cowley, Barry Cox, John Craigie,

Acknowledgements

Len Davis, Vic Davis, Michelle Deas, John Dohle, Karen Dunlop, Charles Edbrooke, Mae Evans, Russell Gray, Shane Hancock, Shane Heck, Joel Hill, Nadia and Cameron Hughes, Bradley Jensen, Ben Karreman, Dave Keller, Russell and Diana Ladbroke, Jack Lewis, Bruce Lord, David McConnel, Col McKay, Alison Philp, Mick Self, Trevor Speis, David Swygart, Ray Richards, Tim Richard, Amanda Taylor, Archie Taylor, Mark Wheildon, Tom and Ann Wilkinson, and Fiona Williamson.

Thanks to curators Ellie Buttrose and Michael Hawker at Queensland Art Gallery/Gallery of Modern Art; Julie McLellan and Suzi Moore at Healthy Land & Water; Mayor Graeme Lehmann and Councillors Cheryl Gaedtke and Jason Wendt at Somerset Shire Council; the Museum of Brisbane; Mike Foster, Rob Drury and Dan Sedunary at SEQ Water; Eva Abal of the River Foundation; Jill Rogers, QUT Digital Collections Librarian; the enormously helpful librarians of the John Oxley Library, and Garry Fitzgerald of the Somerset Wivenhoe Fish Stocking Association.

My deep gratitude to the book's early readers Alisa Cleary, Steve Foley, Gwenn Murray, Amanda Niehaus, Mary Philip, Sally Piper and Luke Stegemann.

My sincere thanks to Elizabeth Cowell for her editorial guidance. To the UQP team: thank you for your brilliance and the depth of your support, including Jean Smith and editor Jacqueline Blanchard and publisher Madonna Duffy who started guiding the journey of this book before its first step.

All errors and omissions are my own and, I hope, forgivable.

And, of course, to Alisa – thank you again, thank you always.

Select Bibliography

Anderson, J 1978, *Tirra Lirra By the River*, Penguin, Ringwood.

—— 2012, *The Commandant*, Text, Melbourne.

Warrington, J (ed and translation) 1956, *Aristotle's Metaphysics*, Dent, London.

Baker, JA 2017, *The Peregrine*, Harper Collins, London.

Baker, VR 2013, 'Sinuous Rivers', *PNAS*, vol. 110, no. 21, viewed 14 March 2024, <https://www.pnas.org/doi/10.1073/pnas.1306619110>

Basho (translation Britton, D) 1974, *Narrow Road to a Far Province*, Kodansha, Tokyo.

Beer, A 2022, *The Flow: Rivers, Water and Wildness*, Bloomsbury, London.

Bell, JP 1950, *Moreton Bay and How to Fathom It: The Yachtsman's Guide*, Queensland Newspapers, Brisbane.

Berger, J 1985, 'The White Bird', *The Sense of Sight*, Vintage International, New York.

—— 2016, *Confabulations*, Random House, London.

Berry, W 2019, 'The Rise', *What I Stand On: The Collected Essays of Wendell Berry 1969–2017*, Library of America, New York.

Blair, R 2011, 'Brisbane River Poetry', *Fryer Folios,* vol. 6, no. 1, University of Queensland's Fryer Library, Brisbane.

Bottoms, T 2013, *Conspiracy of Silence: Queensland's Frontier Killings*, Allen & Unwin, Sydney.

Burger, A 1979, *Neville Bonner: A biography*, Macmillan, Melbourne.

Campanella, R 2010, *Lincoln in New Orleans: The 1828–1831 Flatboat Voyages and Their Place in New Orleans History*, Lafayette Press, Lafayette.

Chambers, P 2004, 'The Origin of Harriet', *New Scientist*, vol. 183, issue 2464, Daily Mail and General Trust, London.

Christopher, E 2022, '"Not a Kanaka or a N____": Reading Pacific labour trading through the slave pasts of sailors of African origin', *Australian Journal of Biography and History*, no. 6, ANU Press, Canberra.

Clark, C 2018, 'Water justice struggles as a process of communing', *Community Development Journal*, vol. 54, no. 1, Oxford Univeristy Press, Oxford.

Clark, C & Page, J 2019, 'Of Protest, The Commons, and Customary Public Rights: An Ancient Tale of the Lawful Forest', *UNSW Law Journal*, vol. 42(1), UNSW Law School, Sydney.

Clark, C, Emmanouil, N, Page, J & Pelizzon, A 2018, 'Can You Hear the Rivers Sing? Legal Personhood, Ontology, and the Nitty Gritty of Goveranance', *Ecology Law Quarterly*, vol. 45:787, UC Berkeley School of Law, California.

Clarke, AR & Zalucki, MP 2004 'Monarchs in Australia: On the winds of a storm?', *Biological Invasions*, vol. 6, Springer Science + Business Media, Berlin.

Condon, M 2010, *Brisbane,* New South Press, Sydney.

Cook, M 2019, *A River With a City Problem: A History of Brisbane Floods*, Univeristy of Queensland Press, Brisbane.

——— 2022, 'Australia's Entanglement with Global Cotton', *Agricultural History*, vol. 96, issue 1–2, Duke Univeristy Press, Durham.

Cook, M & Lloyd, A 2018, 'Unpacking a Legend', *Circa, The Journal of Professional Historians*, Issue 6, Professional Historians Australia, Canberra.

Cranfield, L 1960 'Life of Captain Patrick Logan', *Royal Historical Society of Queensland Journal*, vol. 6, no. 2, Royal Historical Society of Queensland, Brisbane.

Dalaipi, 1958–59, Letters to the Editor of the *Moreton Bay Courier,* 16 November 1858, 24 November 1858, 11 December 1858, 8 January 1859 & 26 January 1859, viewed on 22 March, <trove.nla.gov.au>.

Darwin, C 2011, *Journal of Researches into the Natural History and Geology of the Countries Visited during the Voyage of HMS Beagle round the World, under the Command of Capt. Fitz Roy, RN,* Cambridge University Press, Cambridge.

Day, G 2022, *Words are Eagles: Selected Writings on the Nature & Language of Place*, Upswell, Perth.

DeLacey, E 2016, 'Legends of the Brisbane River Valley', Brisbane Valley Heritage Trails, Brisbane.

Denham, P 2014, *The River: A History of Brisbane,* Museum of Brisbane, Brisbane.

Dixon, P & Millar, D 2004, *150 Years of Brisbane River Housing*, Dixon Partners, 2004.

Dowling, RM 2011, 'Report on the effects of the January 2011 flood on the mangrove communities along the Brisbane River', Queensland Department of Environment and Resource Management, Brisbane.

Erskine, W 1996, 'Environmental Impacts of Tidal Dredging on the Brisbane River, Queensland', *Proceedings of First National Conference of Stream Management in Australia*, vol 1, viewed 14 March 2024, <https://www.researchgate.net/publication/235247162_Environmental_impacts_of_tidal_dredging_on_the_Brisbane_River_Queensland>.

Finnane, M 2008, 'Wolston Park Hospital, 1865–2001: A retrospect', *Queensland Review*, vol. 15, no. 2, Cambridge Univeristy Press, viewed 14 March 2024, <https://journal.equinoxpub.com/QRE/article/view/22098>.

Select Bibliography

Friedel, MH 2020, 'Unwelcome guests: A selective history of weed introductions to arid and semi-arid Australia', *Australian Journal of Botany*, vol. 68, CSIRO, viewed 14 March 2024, <https://www.publish.csiro.au/bt/BT20030>

Gammage, B 2011, *The Biggest Estate on Earth: How Aborigines Made Australia*, Allen & Unwin, Sydney.

Hall, J, Gillieson, DS & Hiscock, P 1988, 'Platypus Rockshelter (KB:A70), SE Queensland: Stratigraphy, Chronology and Site Formation', *Queensland Archaeological Research*, vol. 5, viewed 14 March 2024, < https://journals.jcu.edu.au/qar/article/view/158>.

Harrison, J 2023, *Fettered Frontier: Founding the Moreton Bay Penal Settlement, The First Four Years 1822–1826*, Boolarong Press, Brisbane.

Hesse, H 1954, *Siddhartha*, Picador, London.

Hoskins, I 2020, *Rivers: The Lifeblood of Australia*, NLA Publishing, Canberra.

Hughes, R 1987, *The Fatal Shore*, Collins Harvill, London.

Kenway, S 2013, 'The Water-Energy Nexus and Urban Metabolism – Connections in Cities', Urban Water Security Research Alliance Technical Report no. 100, Urban Water Security Research Alliance, Brisbane.

Kenway, SJ, Renouf, M, Allan, J, Tarakemehzadeh, N, Moravej, M, Sochacka, B, & Surendran, M 2022, 'Urban metabolism and Water Sensitive Cities governance: Designing and evaluating water-secure, resilient, sustainable, liveable cities', *Routledge Handbook of Urban Water Governance*, Routledge, Oxfordshire.

Kerkhove, R 2020, 'Mouth of the Brisbane River: Aboriginal History', viewed March 2024, <https://www.academia.edu/47897756/Mouth_of_the_Brisbane_River_Aboriginal_History_Report_for_Creative_Move>.

Kerkhove, R & Uhr, F 2019, *The Battle of One Tree Hill*, Boolarong, Brisbane.

Kerr, RS 1988, *Confidence and Tradition: A History of the Esk Shire*, Council of the Shire of Esk, Esk.

Kidd, R 2000, 'Aboriginal History of the Princess Alexandra Hospital Site', Diamantina Health Care Museum Association Inc., Brisbane.

Ladd, M 2012, *Karrawirra Parri: Walking the Torrens from source to sea*, Wakefield Press, Adelaide.

Laing, O 2011, *To The River*, Canongate, London.

Lang, JD 1861, *Queensland, Australia: A Highly Eligible Field for Emigration and the Future Cotton Field of Great Britain*, Edward Stanford, London, viewed 14 March 2024, <https://espace.library.uq.edu.au/view/UQ:216484>.

Langevad, G 1979, 'Captain Coley – Queensland's First Sergeant-At-Arms', *Royal Historical Society of Queensland Journal*, vol. 10, no. 4, Royal Historical Society of Queensland, Brisbane.

Lockyer, N 1919, 'Exploration by Edmund Lockyer of the Brisbane River in 1825 – Read before the Society on 28 April 1919', viewed 14 March 2024, <espace.library.uq.edu.au, s18378366_1920_2_1_54>.

Lucashenko, M 2023, *Edenglassie*, University of Queensland Press, Brisbane.

Luino, F, Tosatti, G, Bonaria, V 2014, 'Dam Failures in the 20th Century: Nearly 1,000 Avoidable Victims in Italy Alone', *Journal of Science and Engineering A3*, vol. 3, no. 1, viewed 14 March 2024, <https://www.researchgate.net/publication/264540093_Dam_Failures_in_the_20th_Century_Nearly_1000_Avoidable_Victims_in_Italy_Alone>.

Marshall, T 2015, *Prisoners of Geography: Ten Maps that Tell You Everything You Need to Know About Global Politics*, Elliot and Thompson, London.

Meston, A 1892, 'The Bunya Mountains', *Queenslander*, 21 May, viewed 8 March 2023, <https://trove.nla.gov.au/newspaper/article/3541560>.

—— 1901, 'Name of the Brisbane River', *Queenslander*, 17 August.

—— 1903, 'Morton Bay and Islands – II.' *Queenslander*, 17 October, viewed 18 March 2024, <https://trove.nla.gov.au/newspaper/article/21650486>.

Meyer, A, Schloissnig, S & Franchini, P et al. 2021, 'Giant lungfish genome elucidates the conquest of land by vertebrates', *Nature,* issue 590, viewed 18 March 2024, <https://www.nature.com/articles/s41586-021-03198-8>.

Muir, J 1911, *My First Summer in the Sierra*, Houghton Mifflin, Boston.

Nicholls, H 2006, 'Tall Tales and Tortoises', *New Scientist*, vol. 191, Issue 2560, Daily Mail and General Trust, London.

Oxley, J 1823, *Fieldbooks,* State Library of Queensland, John Oxley Library holdings, Brisbane.

Pearce, F 2018, *When the Rivers Run Dry: The Global Water Crisis and How to Solve It*, Granta, London.

Perera, D, Smakhtin, V, Williams, S, North, T & Curry, A 2021, 'Ageing Water Storage Infrastructure: An Emerging Global Risk' *UNU-INWEH Report Series*, issue 11, United Nations University Institute for Water, Environment and Health, Hamilton, Canada.

Pelizzon, A et al. 2021, 'Yoongoorrookoo: The emergence of ancestral personhood', *Griffith Law Review*, Taylor & Francis, London, viewed 18 March 2024, <https://www.tandfonline.com/doi/full/10.1080/10383441.2021.1996882>

Petrie, C 1904, *Tom Petrie's Reminiscences of Early Queensland*, Watson, Ferguson & Co, Brisbane.

Popelka, SJ & Smith LC 2020, 'Rivers as political borders: a new subnational geospatial dataset', *Water Policy*, vol. 22, issue 3, IWA Publishing, London.

Queensland Floods Commission of Inquiry Final Report, March 2012, Viewed 18 March 2024, <http://www.floodcommission.qld.gov.au/publications/final-report/>

Richards, J, 'Historical changes of the lower Brisbane River' in Tibbetts, IR, Rothlisberg, PC, Neil, DT, Homburg, TA, Brewer, DT, & Arthington, AH (eds) 2019, *Moreton Bay Quandamooka & Catchment: Past, present, and future*, Moreton Bay Foundation, Brisbane.

Shapcott, T 2011, 'The River at Brisbane', *Fryer Folios,* vol. 6, no. 1, University of Queensland's Fryer Library, Brisbane.

Shepherd, N 2014, *The Living Mountain*, Canongate, London.

Select Bibliography

Steele, JG 1970, 'Pamphlet, Uniacke and Field', *Queensland Heritage vol. 2, issue* 3, Oxley Memorial Library Advisory Committee for the Library Board of Queensland, 1970.

—— 1972, *The Explorers of the Moreton Bay District (1770–1830)*, University of Queensland Press, Brisbane.

—— 1975 *Brisbane Town in Convict Days 1824–1842*, University of Queensland Press, Brisbane.

Steffensen, V 2020, *Fire Country: How Indigenous Fire Management Could Help Save Australia*, Hardie Grant, Melbourne.

Suzuki, D 1997, *The Sacred Balance*, Allen & Unwin, Sydney.

Terry, L & Prangnell, J 2009, 'Caboonbah Homestead "Big Rock" or "Little Britain": A study of Britishness in late 19th and early 20th century rural Queensland,' *Australasian Historical Archaeology*, vol. 27, The Australasian Society for Historical Archaeology, Sydney.

Thomson, S, Irwin, S & Irwin, T 1998, 'Harriet, The Galapagos Tortoise: Disclosing One and a Half Centuries of History', *Reptilia*, no. 2, Barcelona.

Wagan Watson, S 2004, *Smoke Encrypted Whispers,* Univeristy of Queensland Press, Brisbane.

Ward, M 2011, *The Comfort of Water: A river pilgrimage,* Transit Lounge, Melbourne.

Watson, FJ 1944, 'Vocabularies of four representative tribes of South Eastern Queensland', Queensland: Royal Geographical Society of Australasia, Brisbane.

Willmott, WF, & Stevens, NC 1992, 'Rocks and Landscapes of Brisbane and Ipswich: Geology and Excursions in the Brisbane, Ipswich and Pine Rivers Districts', Geological Society of Australia, Queensland Division, Brisbane.

Artwork

Sam Cranstoun, *A Simple Story*, 2022

Michael Parekowhai, *The World Turns*, 2011–12

Richard Tipping, *Watermark*, 2000

Anne Wallace, *Passing the River at Woogaroo Reach*, 2015

Miscellaneous

Cope, M 2023, interview 'The Black Napoleon', viewed 4 November 2023, <https://mosmanartgallery.org.au/exhibitions/megan-cope-black-napoleon-re-formation-i>

'Carving A History: A Guide to the Great Court, The University of Queensland, St Lucia' 1979, The University of Queensland, Brisbane.

Hydria Virtual Museum: Linking Ancient Wisdom to Modern Needs, viewed 4 November 2023, <https://hydriaproject.info/en>

Court Decisions

Edith Pastoral Company Pty Ltd v Somerset Regional Council & Ors, 2021, QPEC 52

Murphy on behalf of the Jinibara People v State of Queensland, 2012, FCA 1285

Queensland Bulk Water Supply Authority t/as Seqwater v Rodriguez & Sons Pty Ltd, 2021, NSWCA 206

Sierra Club v Morton, 1972, 405 US 727

Legislation

Land Act 1994 (Qld)

Te Awa Tupua (Whanganui River Claims Settlement) Act 2017 (NZ)

Water Act 2000 (Qld)

Yarra River Protection (Wilip-gin Birrarung murron) Act 2017 (Vic)